Praise for
Here Be Monsters

'Endlessly fascinating. An extraordinary inquiry into the hidden ways in which technology shapes and reshapes human beings and our world, by one of our most stylish and elegant writers.'

Guy Rundle, journalist and critic

'Like a cave-diver, Richard King steadily explores his way through the chambers of consequence that lie beneath, around, above and within our relationship to technology. Some are easily illuminated; others keep their shadows, but King sounds out the dimensions, the contours and the crags of this world in which human and machine are together becoming more and more submersed in unknown waters.

'Concerned and sceptical but not unjust, King surveys both the big innovations and the philosophical legacies of this tech age, somehow finding space for meditations on humanity, an astute grasp of upcoming invention, and the posing of fierce, urgent questions. It is, he says, "humanity's ability to *ask* what is suitable – what is good, what is bad, what is progress, what is regress – that separates it from other species". In this excellent, very readable and laudably ambitious work, Richard King takes nothing for granted, but gives us a portrait of a species in the act of utterly changing itself, a terrible beauty being born.'

Kate Holden, journalist and author of the Walkley Award–winning *The Winter Road*

'Technologies like artificial intelligence are changing our world. But all too often, technology is seen as destiny. *Here Be Monsters* is an important and engaging look at how these tools are using us, and how we must act to regain our essential humanity.'

Toby Walsh, chief scientist at UNSW AI Institute and author of *Machines Behaving Badly*

T0359639

'*Here Be Monsters* is an intelligent and thoughtful meditation on the relationship between technology and humanity. Pulling together tech criticism, literary theory and history, King has created a text that is bigger than the sum of its parts. This thoroughly enjoyable book gives the reader the confidence to commit to a bold ambition for a more democratic technological future.'

Lizzie O'Shea, lawyer, activist and author of *Future Histories* and *Empowering Women*

Praise for *On Offence*

'Magnificent'
The Observer (UK)

'One of the books of the year ... an intellectual map for our times.'
The Saturday Age

'An extended essay of uncommon eloquence and brio.'
The Australian

'A calm, clever and lucid book.'
Gideon Haigh

'Lively ... a bright and magnanimous reminder.'
The Sunday Age

'A ripper of a book ... Fantastic.'
ABC Radio

'Should be required reading'
The New Zealand Listener

'Fired me up'
Andrea Goldsmith, author of *The Memory Trap*

HERE BE MONSTERS

Is technology reducing our humanity?

Richard King

MONASH
UNIVERSITY
PUBLISHING

For Sarah, Tom and Lucy

Published by Monash University Publishing
Matheson Library Annexe
40 Exhibition Walk
Monash University
Clayton, Victoria 3800, Australia
publishing.monash.edu

Monash University Publishing: the discussion starts here

9781922633385 (paperback)
9781922633408 (epub)

Cover design by Peter Long
Author photograph by Bohdan Warchomij

Printed in Australia by Griffin Press

[Man] aspires to make robots as one of the greatest achievements of his technical mind, and some specialists assure us that the robot will hardly be distinguished from living men. This achievement will not seem so astonishing when man himself is hardly distinguishable from a robot.

ERICH FROMM

Where is the Life we have lost in living?
Where is the wisdom we have lost in knowledge?
Where is the knowledge we have lost in information?

T. S. ELIOT

CONTENTS

CONTENTS

PROLOGUE

This book, which has the word 'monsters' in its title, is itself a kind of monster: a chimera. Like the mythological creature composed of different animal parts, it doesn't fit into a single category. It is a book about technology, but it is not written from a 'techy' perspective. It is written from the perspective of one trained in the humanities – in literature principally, and, via literature, in philosophy and cultural criticism. That's to say, it is a book of ideas, on a subject – technology – often taken to sit substantially outside the realm of ideas. Moreover, it is a book of ideas in which very different disciplines – psychology, anthropology, political economy, the arts – are often to be found in one another's company. It is, in short, something of a hybrid, albeit one with a firm perspective and (I like to think) a bit of fire in its belly.

There are countless books on technology written by the adepts of artificial intelligence, nanotechnology, modern genomics and so on, some of them indispensable. But as technology, and the human *relationship* to technology, enters a revolutionary new era, I'm convinced we need to go beyond the prescriptions of the technology 'community' and come at the issue from a different angle. Not to do so, on the basis that one is not qualified to comment on the latest interventions, is to concede the nature of the issue in advance, and to accept the instrumental framing that so often accompanies technological matters. We need to ask what we *want* from technology. And the question of what we want is a question about who we *are*.

In essence, then, this is a book about us. As the boys and girls (mostly boys) of Silicon Valley never tire of telling us, human beings are tool-using animals. But this doesn't mean that every tool we use is good for our humanity. To believe that is to place technology beyond the sphere of politics, which is the word we give to the study and practice of living in community with others. We need to draw it back into that sphere – to ask not only how we are using our tools but also how our tools are using us.

In order to address the many complex ways in which new and emerging technologies work against the human condition, I have divided the text into three sections, each of which corresponds to what I take to be an essential facet of human nature: the social, the physical, and the purposive and creative. As readers will quickly discover, however, these different facets only really make sense when considered in combination with one another: for example, one cannot begin to understand the way we live together, and what forms of society are best for us, without also recognising that human sociality evolved in conditions of physical co-presence – through face-to-face contact. The point is worth underlining, I think, because the modern 'technosciences' (the fusion of science and technology under pressure from the profit motive) have fostered a simplistic idea of human nature. I wrote this book in order to analyse and challenge that view. Human beings are far more complex and fascinating than any simple list of characteristics can suggest, and my three-part structure aims to emphasise that, not to imply reductionism.

This book is part diagnosis, part prognosis, part symptomatology and part prescription. Often, when books meet the mainstream media, the emphasis will tend to fall on the last thing – on what we should do about the problems identified. I have some things to say about that, but above all I want to emphasise that what we need now is a new way of thinking about technology – new habits of mind, or the 'power of facing', as George Orwell famously described his own gift. In the end, there is no hard-and-fast distinction between analysis and action, or calls for action. Both are aspects of the same adventure, which is the adventure of a uniquely gifted species in control of its own destiny.

The conundrums and examples in this book have a general relevance, bearing as they do on the relationship between technology and human nature. Nevertheless, in terms of their technological character, the Global North and Global South (the rich world and the poor world, broadly speaking) are at very different stages of development. It is partly because I want to question one version of 'development' – and its conceptual bodyguards, 'innovation' and 'progress' – that I have chosen to focus on the Global North, whose relationship to technology is now entering a revolutionary stage. That relationship, of course, would not be possible without the exploitation of the Global South, but it is, in the end, the character of that relationship with which I am concerned, not questions of global economic justice. Suffice it to say that I don't believe our own model of development is the only one possible, and that those who talk about economic justice as if all the Global South has to do is to emulate the Global North are very much part of the problem, as I see it. For all its material advantages, the North has problems the South doesn't have, or has to a lesser degree – problems for which the figure of the monster will, for the next 200 pages or so, function as a metaphor.

INTRODUCTION

Night of the Living AI

In 1968 two movies were released that, although different in many respects, had one fundamental thing in common: they prompted us to reflect on the kind of creatures we are by showing us the kind of creatures we are not.

George Romero's *Night of the Living Dead* is a brutal, black-and-white, low-budget picture, widely regarded as a watershed in the development of modern horror. Often cited as an example of 'exploitation cinema', and as the progenitor of the 'splatter film' subgenre, its director sought to capitalise on the movie industry's thirst for the bizarre with a level of terror and violence unseen in horror movies up to that point. (Legend has it that the gory effects were achieved with Bosco Chocolate Syrup and roast ham, plus offal from a butcher's shop owned by one of the cast members.) But it wasn't just the violence that caught the movie-going public's attention. It was also the creature at the centre of the story: a chaotic, shambling revenant with a frozen stare and tattered garments, a rigor-mortis-stiffened gait and an insatiable craving for human flesh. Staggering through *Night of the Living Dead* was an army of what we now term 'zombies'.

The origins of the zombie lie in Haitian folklore. Derived from African folk traditions and imported into Haiti through the Atlantic slave trade, the 'zombi' was a figure summoned from the dead by a necromancer, or sorcerer, who would then control its actions through magic. But it was Romero's re-imagining of the zombie that gave us the archetype we recognise today, his innovation being to combine it with the essence-sapping figure of the European vampire. *Night of the Living Dead* features an army of ghouls reanimated not by voodoo magic but by radiation from a crashed satellite, and bent on gaining entry to a farmhouse and feasting on its human residents.

Spoiler alert: it doesn't end well.

Stanley Kubrick's *2001: A Space Odyssey* is a lavish science-fiction epic – a slow, hypnotic visual ballet with minimal dialogue and a classical score, esteemed for its groundbreaking special effects and realistic depictions of spaceflight. Set aboard the *Discovery One* spacecraft, it centres on the relationship between the ship's computer, HAL 9000, and the only two astronauts not in hibernation: Dave Bowman and Frank Poole. HAL is an Artificial Intelligence, or rather an Artificial *General* Intelligence – a sentient computer with the ability to understand or learn any intellectual task a human can. In a human voice, he interacts with Bowman and Poole in a friendly, if affectless, way. But as the film progresses he begins to malfunction, and the 'eyes' through which he keeps tabs on the ship – intense orange dots with red penumbras – take on an ominous quality. Convinced that the astronauts are planning to disconnect him, HAL (spoiler) sets about killing the crew, before Bowman makes his way to the processing core and removes the computer's cognitive circuits. HAL's final moments are strangely moving. As the AI's cognitive abilities degrade, it falls back on its earliest programs and begins to sing an old music-hall ditty, much as a human being might do once senility has taken hold.

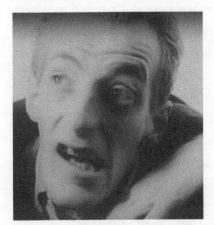 

The first modern zombie, from *Night of the Living Dead*, and HAL 9000, from *2001: A Space Odyssey*

Both HAL and Romero's zombies are monsters: creatures that fall outside, or between, conventional categories. Like centaurs, werewolves, vampires or spirits, they occupy an ambiguous territory between different states or physical forms, in a way that subverts, or threatens to subvert, the social, moral or natural order. Aristotle, who believed that every creature had its own fundamental *telos*, or purpose, saw monsters as mistakes of nature: their development had been corrupted such that they were unable to become the thing that nature or God intended. Sometimes described as 'liminal' entities (from the Latin word for threshold, *limen*), monsters are personifications of what Sigmund Freud described as 'the uncanny': a sort of emotional dissonance by which figures can appear at once familiar and foreign. Severed limbs that move independently of their bodies, incorporeal spirits and evil doppelgangers are just a few examples of this phenomenon, which Freud, being Freud, explained in terms of ego, narcissism and unconscious impulses. Penetrating to some visceral core, such liminal beings induce feelings of anxiety, queasiness and outright disgust.

The liminality of the zombie is easy to identify. For here the monster is neither alive nor dead, but something in between the two: 'the living dead' or 'the walking dead' or simply 'the undead'. Sometimes it is 'the evil dead', but its evil character is secondary to its ambiguous status as life without being. As in the folklore from which it derives, there is a dualism in our conception of the zombie, a sense that the soul or spirit of the human being is detachable from its material form, and that its disappearance leaves us at the mercy of our basest animal appetites. The fact that zombies crave human flesh is the sign that they are no longer human, and yet it is their similarity to humans that induces such anxiety in the spectator. They exist in the 'uncanny valley' between the barely human and the fully human.[1] Even before it begins chowing down on the poor old widow in the rural farmhouse, the dead-eyed zombie deflects human empathy. To be 'almost human' is, it seems, to be creepier than not being human at all.

HAL represents a different archetype, and a different kind of liminality: the killer AI that threatens to surpass, and eventually overrule, its human creators. He is moving towards humanness, rather than away from it, in a way that channels longstanding anxieties about the power and potential of new technologies. As such, he belongs in an alternative tradition of entities that arouse our anxiety because they appear to be *gaining* on us. What makes them uncanny is the ambiguous character of their intelligence and morality. At times in *2001*, for example, HAL appears paranoid and even a little jealous. There is a suggestion that his apparent malfunctions are in fact a sort of mental breakdown, or even an existential crisis brought on by the evidence of his own fallibility, or perhaps by the stress of having to conceal the truth of *Discovery One*'s mission from its crew. 'I'm afraid, Dave,' he confides to Bowman as his cognitive functions deteriorate. 'My mind is going. I can feel it. I can feel it.' But even a computer of HAL's complexity shouldn't be able to *feel* anything at all.

So, two very different monsters, channelling very different fears: one focused on our degeneration into something less than human, one on the increasing sophistication and power of new technologies.

But here's the question I can't get out of my head. What if there was a third scenario in which these anxieties came together? What if the danger of new technologies lay not in their power to *subjugate* human beings but in their power to *transform* us into something we were not? And what if, having been so transformed, we became uncanny to ourselves and one another?

What might the monsters look like then?

Humanity at the Threshold

The co-discoverer of DNA, James Watson, likens it to the *Apollo 11* mission that put human beings on the moon for the first time. Others compare it to the splitting of the atom and the development of the first atomic weapon. These comparisons are reasonable: there is no question that one of the most important moments in the history of science and technology was the announcement that scientists had 'mapped' the human genome.

The announcement was made by Bill Clinton in his last year as US President, on 26 June 2000. Flanked by the geneticist Francis Collins and the biotechnologist Craig Venter – leaders, respectively, of the Human Genome Project and the Celera Corporation, whose creative rivalry had ensured that the genome 'draft' came in well ahead of schedule – Clinton mounted the podium and, after acknowledging a few dignitaries, addressed the assembled media:

> Nearly two centuries ago, in this room, on this floor, Thomas Jefferson and a trusted aide spread out a magnificent map, a map Jefferson had long prayed he would get to see in his lifetime.
>
> The aide was Meriwether Lewis and the map was the product of his courageous expedition across the American frontier all the way to the Pacific. It was a map that defined the contours and forever expanded the frontiers of our continent and our imagination.
>
> Today the world is joining us here in the East Room to behold the map of even greater significance. We are here to celebrate the completion of the first survey of the entire human genome. Without a doubt, this is the most important, most wondrous map ever produced by humankind.[2]

Clinton's comparison between the 'mapping' of the human genome and the mapping of the newly expanded United States by Lewis and Clark was striking and germane. But it was also full of resonances that Clinton presumably didn't intend. For Lewis and Clark's expedition across North America was first and foremost a commercial enterprise, undertaken by representatives of the United States' incipient empire. Instructed by Jefferson to declare 'sovereignty' over the various Native American tribes they encountered along the Missouri River, Lewis and Clark were part of a history that began with the Spanish colonisation of the Americas in the late fifteenth century and encompassed countless massacres and displacements, as well as the enslavement of native populations, war and the destruction of traditional ways of life. In other words, they were part of a history in which 'inferior' societies were forced to make way for the more 'advanced' European ones.

Similarly, the Human Genome Project's sequencing of human DNA did not occur in an historical vacuum, and many commentators have raised the possibility that recent developments in genetic technology could reproduce or even deepen existing prejudices and inequalities. The sci-fi movie *Gattaca*, which imagines a society based on eugenics (or 'artificial selection'), is a dystopian fantasy. But the issues it raises are entirely pertinent to the development of genetic engineering, especially in light of the recent emergence of CRISPR/Cas9 technology, which enables scientists to 'edit' an organism's genome by replacing strips of its DNA. This technology will make it possible to eliminate heritable diseases such as sickle cell anaemia, but it also brings 'designer babies' and other cosmetic interventions into prospect. I don't think it's hysterical to imagine the emergence of a new biological class system mirroring the economic one – a caste of genetically 'superior' people, more beautiful or more intelligent than the rest of us. Moreover, and perhaps less obviously, such interventions could change the way we interact with one another, altering the texture of the social fabric (the parent–child relationship; our notions of talent, luck and equality). Geneticists talk of 'off-target effects', in which specific genetic interventions cause unintended mutations or deletions. But surely the *social* off-target effects of genetic engineering will be far more profound.

In all kinds of ways, then, we are at a threshold. If 'genomics' is a map, as Clinton claimed, it is one on which the cartographer's warning *Hic Sunt Dracones* appears at the margins: 'Here Be Dragons' ... 'Here Be Monsters'.

In Virgin Territory

What goes for genetic engineering goes too for other recent developments. Biotechnology, nanotechnology and information technology have expanded the frontiers of the human continent, and we are only beginning to learn what riches and dangers lurk in its alien terrain. One thing we *do* know for certain: we will soon be living in a very different world.

To some extent this has always been so. Human beings are technological animals, with the ability to make tools, and tools to make tools. These tools are used to transform the environment, which, over time, and according to the same processes of Darwinian evolution that shape all animals, eventually changes human beings. Human beings and technology, then, co-evolved in a virtuous spiral of 'enhancement'. As far as we know, the use of stone tools predates *Homo sapiens* by some three million years – such that when we did emerge as a species distinct from our hominid ancestors, the ability to use technology was innate. And not just innate, but key to our survival: without the dexterity and practical understanding needed to use technologies, *Homo sapiens* could not function in the world. We are *Homo faber*: 'man the maker'. No less than the instinct for self-preservation or sexual desire, technological creativity is fundamental to our being.

Take the ability to manipulate fire. That discovery, which would have occurred in different places and over millennia, provided our ancestors with a source of warmth, saving precious energy and extending the range of habitable territories. It also allowed for the cooking of food, which breaks down fibres and plant-cell walls in what amounts to a form of pre-digestion. This too saved precious energy, which could be used to run a larger brain capable of creating *new* technologies – garments, shelters, hunting weapons – that increased our energy 'efficiency' still further. Thus humans manipulated fire, and fire manipulated humans. We are both the producers and the products of our tools.

But today we stand in a radically different relationship with our environment. We have moved from simply manipulating nature to re-constituting its smallest elements.

This change is not merely a matter of degree. It is a revolution in our relationship with the world. For we are able not only to harness capricious nature according to our will, but also to *intervene* in nature at the level of the atom, the molecule and the gene. We can now intervene in the very *stuff* of nature itself, up to and including our own biology.

Perhaps the most dramatic moment in this trajectory was the splitting of the atom at Los Alamos on 16 July 1945, in the first-ever test of an atomic weapon. Forty years earlier, in 1905, Einstein had theorised the equivalence of mass and energy in his equation $E=mc^2$, which said that under certain conditions energy becomes mass and vice versa, and that the amount of energy contained within mass was such that even a tiny object would cause a catastrophic explosion in the event of a mass-to-energy conversion. But for the young theoretical physicist there was no practical way to convert such an object from mass into energy to test the theory. It was only when the scientists of the Manhattan Project, who were working to develop an atomic weapon towards the end of World War II, found a way to release the energy stored within uranium and plutonium (both 'unstable' elements) that his theory found practical application – an application that demonstrated human beings could not only utilise nature but *convert* it, and led in just a few short weeks to the bombing of Hiroshima and Nagasaki and the deaths of over 200,000 people.

The Trinity Test at Las Alamos, 0.016 seconds after detonation

Technologically this was virgin territory indeed. Nuclear power transformed the world at the level of the atom, moving scientific endeavour from understanding to authorship. As William L. Laurence, the only journalist present at the test site in Los Alamos, intuited, the blast represented a radical break. It was 'the first fire ever made on Earth that did not have its origin in the Sun', he wrote.[3]

The development of the atomic bomb inaugurated the broader shift from manipulation to reconstitution that is now taking place across a range of fields. Nanotechnology means we can fabricate new materials on the scale of atoms and molecules. Emerging biotechnologies are poised to enhance human longevity by turning back the telomere 'clock' that triggers cells' demise. Information technologies allow tech companies not only to seed the world with new forms of intelligence but also to harvest huge amounts of data with a view to constructing models of individual internet users – a voodoo-doll vision of humanity in which people are perpetually dosed, nudged and prodded with subliminal cues. In this way, and in others, new technologies promise (and threaten) to *remake* 'man the maker'.

And these fields are combining in ways likely to accelerate the development of each. Nanotechnology allows scientists to apply their own specifications to living cells, changing the building blocks of organic life. Strides in genetic sequencing would have been impossible without comparable advances in information technology. In a breathless report published in 2002, 'Converging Technologies for Improving Human Performance', the US National Science Foundation proposed an acronym for this 'technological convergence': NBIC (Nanotechnology, Biotechnology, Information technology and Cognitive science). Much of the report is fanciful, speculating on a future of 'world peace, universal prosperity, and evolution to a higher level of compassion and accomplishment', as well as on the potential for 'brain-to-brain interaction'. There is also a reference to something called 'the Communicator', which will 'remove barriers to communication caused by physical disabilities, language differences, geographic distance, and variations in knowledge'.[4] But there is no doubt the authors were onto something. One of the report's contributors was so taken with the notion of convergence that he penned a poem:

If the *Cognitive Scientists* can think it
The *Nano* people can build it
The *Bio* people can implement it, and
The *IT* people can monitor and control it

T. S. Eliot it isn't, but as a description of the way transformative technologies have combined to become more transformative still, this is pretty accurate.

The Fetish of Progress

Whereas science was once primarily concerned with exploring the physical environment through observation and experimentation, in ways that may or may not translate into material application in the future, it is now oriented around problem-solving. Science and technology have largely fused, and the traditional scientific distinction between 'pure' and 'applied' knowledge is seldom made. The default objective is often practical utility, not understanding or knowledge for its own sake.

The usual name for this new fusion of science and technology is 'technoscience', and its drivers and characteristics will be discussed throughout this book. For now it is only necessary to note that one effect of its emergence has been to disguise the developments described here, usually under the heading of 'progress'. According to the technoscientific worldview, the new technologies of transformation are no different in principle from the discovery of fire or the invention of the internal combustion engine: humans are just doing what they've always done. Indeed, they are fulfilling their *telos* – being true to the kinds of creatures they are in precisely the way Aristotle envisaged. Whether it's a hammer or a hard drive, an animal fur or an antidepressant, it all comes to the same thing, in the end.

Consequently, the more instrumental attitude to scientific understanding that characterises the technosciences leaves out wider questions about the way we live – about our relationship to our environment and to one another. New technologies are deemed value-neutral: they may be put to terrible ends but they are not terrible in themselves, the argument runs, and to suggest otherwise is to make a kind of category error – to confuse the world of human affairs (politics, the will to power and so on) with the arc of technological progress. 'It isn't social media that's the problem,' it's said, 'but the way in which it's used.' Similarly, most people would readily agree that gene editing may be used unwisely, but few would question the wisdom of developing gene-editing technologies in the first place.

What proponents of this view fail to appreciate, though, are the social, economic and political conditions under which technological innovation occurs. Would social media have been invented at all in a society that accorded greater value to *physical* community life, or one less in thrall to performative individualism? Dominated by a marketplace that fetishises choice and individualism, and necessitates private ownership and profit, we are carried towards goals we had no hand in setting. We are told that progress is being made but not what the destination is.

The fetish of progress also serves to disguise the way these new technologies, developed under particular conditions and according to particular priorities, bring with them a new vision of the human animal. Science's move from the field to the lab towards the end of the nineteenth century led to a general de-emphasis on context and environment. Moreover, it led to a view of life as composed of ever-smaller elements – of atoms and genes – that contained 'information'. With the rise of information theory in the 1940s, and especially the invention of the 'bit' (the basic unit of information used in computing and digital systems), the scene was set for a radical change in our self-perception. Over time we came to see ourselves as intricate, largely autonomous systems, no different from complex machines. We began to see life in *informational* terms. As computers increased in power and complexity, it became natural to think that their 'intellectual' evolution would ultimately intersect with our own. This view is almost certainly erroneous: the whole idea that computer intelligence will reach a point of such complexity that it passes an existential threshold and begins to quibble about who is in charge or cry at the sight of three-legged dogs ignores the fundamental differences between biological and mechanical 'thought'. But its power is there for all to see in the pages of *Wired* and the transhumanist fantasies of Silicon Valley. Even now there are technicians working on ways to 'upload' the human mind to computers.

This ethos of fine manipulation runs right through the technosciences. Mind, body and matter are perceived as informational systems to be modified, customised, recoded or rewritten. Not only is the mind viewed as a more complex version of your MacBook Pro, but genetic engineering invites us to see life in informational terms, as opposed to something intricately marbled into our social and environmental context. Thus new technologies remake the world, while also changing our view of it, and radically new interventions become not just possible but permissible. The more we see computers as sentient beings in prospect, the more likely we are to see ourselves as computers.

This is a crude sketch, but the point is that the approach to life engendered by the technosciences is bringing about a reversal in the 'direction of fit' between human beings and society. Where once we might have regarded, say, a spike in depression and anxiety as an opportunity to consider what could be going wrong in society – in the composition of our communities, in our relationship to nature, in the nature of work – today we are apt to reach for the prescription pad and the latest biochemical preparation, adapting the human body and mind to the societies we have created, rather than building societies in which humans are more likely to flourish. Accepting technoscientific innovation as the measure of our species' progress, we cede responsibility for the future to the reassuring figure in the lab coat and goggles.

The Human Future

The contemporary discussion about new technologies tends to focus on particular issues. Overwhelmingly concerned with digital media, it will often stress the relationship between the Big Tech company or the state on the one hand and the private citizen on the other. What data is collected, by whom and to what end are the principal areas of concern, with the conversation framed in terms of the privacy and rights of individual users. It is a conversation about the relationship between the individual and certain forms of power, and precisely the one we would expect to find in a liberal, rights-based society such as ours.

It's an important conversation, and one I'll touch on in these pages, as we consider the darker implications of digital technologies. It's also one of the principal themes in those dystopian fantasies that explore our fears about advanced technologies. HAL's cyclopean eye taps into contemporary anxieties about the subjugating power of digital technology – about China's 'social credit' system, for example, which uses complex algorithms to assess citizens for their trustworthiness, or about the 'surveillance capitalism' that models internet users with a view to shaping their online behaviour. As the tools we rely on become ever more independent of our physical selves – not extensions of our person, like hammers or spoons, but autonomous systems that are connected to invisible systems elsewhere – issues of control and freedom become crucial. A mere decade ago, social media platforms were invoked as the guarantors of future freedom: tools that would liberate the people of North Africa or the Middle East from repressive regimes. Today we are rather less naive, though no less dependent on our computers and phones.

So, an important conversation to have. But it is, I believe, a limited one, in that it fails to take account of the effects that new technologies will have, and are having, on the fundamental aspects of our humanity. It assumes a private citizen – an individual with rights and responsibilities – that is, ultimately, a political abstraction. It is not on the basis of the human condition that such analyses draw their conclusions but on the basis of an abstract political subject that is itself the product of historical forces.

The issues I want to address, by contrast, relate to our fundamental *humanity*. How are these new technologies, which emerged under the same historical conditions that produced this liberal, rights-based subject, affecting us in our deepest being, and what might this mean for the future of our species? How do technologies that favour relations of absence over relations of presence, or that tempt us to remake ourselves in line with ephemeral notions of success or excellence, affect us on this rudimentary level? Our brains may be artificially extended through books and libraries and the internet, but the human brain is still, substantially, the same that witnessed the last Neanderthals sharpening their axes 50,000 years ago. Can we take it for granted that once we begin tinkering with the building blocks of organic life there won't be some kind of social explosion, more significant even than the one in Los Alamos?

Of course, the question of what human nature is, or whether we even *have* a nature, spurs much disagreement. For some, human nature is an essentialist notion that is liable to lead to discrimination, and one only need look at the way the concept has been used in the past to see they have a point. But we would be making a huge mistake if we allowed this to deter us from talking about human nature at all, for the notion that human culture and identity have nothing to do with human nature is in many ways just as dangerous. The 'blank slate' view of humans – the idea that human identity and behaviour are socially or culturally constructed – has become something of an article of faith in parts of academia. Even in mainstream politics it has substantial currency, especially in discussions of gender and sexuality. But as important as this idea has been in the fight against racism, misogyny and homophobia, it cannot form the basis of a radical critique of emerging and soon-to-emerge technologies. The social constructionists' tendency to treat all talk of human nature as socially and politically regressive is a gift to technoscientific hubris. What better way to justify the steady march of new technologies into every aspect of human existence than by dressing it up in the rainbow colours of individual emancipation!

Even among those who *do* accept that there is such a thing as human nature, there is a wide variety of views about where nature 'stops' and culture 'begins'. But *some* discussion of our limits is essential, and so we need to develop a minimal view of our 'natural–cultural condition' – one that encompasses the social, creative, intentional and physical facets of humanity in a way that stresses not only their centrality but also their deep interconnection. The German philosopher Immanuel Kant wrote of the 'crooked timber of humanity'; but even crooked timber has a grain, and we work against our own at our peril.

The New Prometheans

I began with a question: can we imagine a scenario in which the different anxieties aroused by George Romero's horror film *Night of the Living Dead* and Stanley Kubrick's sci-fi dystopia *2001: A Space Odyssey* merge?

How might a monster that combined our fear of becoming something less than human with our fear of increasingly 'intelligent' machines appear to us, and what might it say?

There is one work – of both horror *and* science fiction – that imagines such a monster. Published almost exactly 150 years before Romero and Kubrick released their movies, it is a book in which physical deformity and technological mutiny coalesce, creating a monster that is both a zombie and AI, or something in between the two. A gothic fiction, it is also described by some literary historians as the first science-fiction novel. Its title is *Frankenstein; or, The Modern Prometheus.*

Mary Shelley's dark Romantic masterpiece was conceived and written on Lake Geneva in the fabled 'year without a summer' – 1816, when volcanic ash from the eruption of Mount Tambora in Indonesia shrouded the Earth – and provides a matchless metaphor for the intersection of science, technology, hubris and short-sighted ambition that characterises the present moment. The titular Victor Frankenstein is a young scientist who develops a secret technique to impart life to non-living matter, and his ambition leads him to use that technique to assemble an entire human being, one anatomical feature at a time, from the bodies of dead humans and animals. Horrified at the results of this experiment, he abandons the newly reanimated creature, whose appearance and size condemn him to a life of lonely, loveless misery. The creature swears revenge on Victor, and pursues him through what remains of the novel, punishing the scientist's friends for his crimes and demanding a companion, upon pain of more carnage. It is a story drenched not in blood and gore but in unbearable longing and desolation. Taking care to piece his creation together so that everything functions as it should, Victor neglects to consider the thing that makes a being fully human: participation in a community that, whatever its injustices and distortions, affords the possibility of acceptance, companionship, understanding and love.

Not unusually for a literary creation that captures the popular imagination, Shelley's sensitive, anguished creature has undergone a significant make-over in its journey from character to archetype. The popular image of

Frankenstein's monster is of a towering, heavy-set, undead-like figure with greenish skin and an angular head. The intelligence and athletic agility of Shelley's creation are no longer in evidence; instead, the monster is as dull and rigid as any post-Romero zombie. The template for this representation is Boris Karloff's interpretation of the monster in James Whale's 1931 movie *Frankenstein*, which also played a major role in transforming another aspect of the Frankenstein story: the character of the monster's creator. For while Shelley's Victor is undoubtedly hubristic, he bears little resemblance to the 'mad scientist' that appears in so many interpretations of the novel. That is a modern characterisation, and a key reason *Frankenstein*, or the Frankenstein story, no longer has the resonance it deserves.

Frankie aka The Creature of Doctor Frankenstein, a cast bronze sculpture in the Plaine de Plainpalais, Geneva, by the KLAT art group

So when a technoscientific innovation – a new genetic intervention, say, in agriculture or medicine – is described by its critics as 'Frankenscience', there tends to be a collective rolling of eyes. Invocations of monstrosity are met in a spirit of amused indulgence. It's almost as if the Frankenstein story, in its extra-literary iterations, is allegorically self-defeating – a warning not against scientific hubris, but against the accusation of such. *Oh, don't be daft,* say the scientists to their critics. *It's just a cheesy fantasy with Herman Munster in the starring role!*

But this overlooks that Shelley's *Frankenstein* is a work of imagination focused not on some single calamity or experimental 'step too far', but on the broader hubris that presumes to treat *being* as a material fact like any other, to be made or modified at will. It is the dramatic encapsulation of a mindset, not the literary equivalent of a disaster movie. Shelley and her literary coterie (which included Lord Byron and her lover, Percy Shelley) were deeply interested in the new techniques emerging from 'natural philosophy': in Benjamin Franklin's experiments with lightning and conductivity, and in the scientific ideas of Erasmus Darwin, whom they believed (erroneously) to have vivified a piece of vermicelli noodle using a technique called 'Galvanism' – the chemical generation of an electrical current. But for Shelley such knowledge cannot be treated in isolation from its human context, and that is why she depicts Victor Frankenstein as the product of an incomplete education – as a man whose analytic tendencies, unanchored in philosophy or the arts, cause him to take a mechanistic and reductive view of humanity. Like the Prometheus with whom he shares the book's title – the Titan trickster who stole fire from the gods and gave it to humanity – he is guilty of insufficient humility in the face of our multifaceted nature. He is not a mad scientist, but a blinkered one.

Looked at in this way, *Frankenstein* is more relevant today than ever. In our technoscientific era, the fundamental elements of nature are manipulated in a spirit of Promethean 'progress', and a reductive and mechanistic idea of humanity is central to that project. The part of Victor is taken by a composite of corporations, governments, the military and the modern university, now largely denuded of its humanistic ethos. And while many of its schemes may turn out to be as fanciful as breathing life into a mouldering cadaver, the mindset that gives rise to those schemes will only go from strength to strength unless we begin, like Shelley, to question the ignorance and arrogance at its core. Some of the interventions entertained in Silicon Valley or the biotech sector would be dangerous if they came to fruition. But more dangerous still is the ideological climate that allows them to be entertained at all.

A Radical Humanism

In his speech on the mapping of the human genome, Bill Clinton reflected:

> Today's announcement represents more than just an epoch-making triumph of science and reason. After all, when Galileo discovered he could use the tools of mathematics and mechanics to understand the motion of celestial bodies, he felt, in the words of one eminent researcher, that he had learned the language in which God created the universe. Today we are learning the language in which God created life. We are gaining ever more awe for the complexity, the beauty, the wonder of God's most divine and sacred gift.

It's a skilful performance – one that tries to allay fears that the technosciences are 'playing God' with a genuflection to The Big Guy Himself, introducing a note of humility into a triumphant moment for science. But there is also hubris lurking in the humility. For as the historian of science Sheila Jasanoff has shown, the idea that genetics constitutes a 'Book of Life' akin to a religious text was quickly taken up by scientists and scientific commentators with no interest in divine revelation.[5] Personally, I can sympathise with their desire to deny the final word on what a human is to a Sky-Daddy tinkering in his cosmic shed. But the idea that genetics alone can encompass what it means to be human is scarcely less ridiculous. Indeed, listen carefully to Clinton's words and one may even hear the ghost of Victor as he contemplates his own endeavours: 'So much has been done, exclaimed the soul of Frankenstein – more, far more, will I achieve; treading in the steps already marked, I will pioneer a new way, explore unknown powers, and unfold to the world the deepest mysteries of creation.'

When the biophysicist He Jiankui took it upon himself to modify the genome of unborn twins in 2018, many scientists objected on medical-ethical grounds. Others called He 'China's Dr Frankenstein'.

I prefer the second response.

_____

Stephen Asma, a professor of philosophy at Columbia College Chicago, writes: 'Our generation is like Dr Frankenstein standing over a table of miscellaneous limbs and organs, only we're on the table, too.' Asma suggests that we interrogate technological developments as incubators of the worldview that allows technoscientific hubris to reproduce and spread. His point is that, to some degree, we are *all* suffering from Victor's delusion, because we are all encouraged to see nature in mechanistic terms, as something that can be bent to our will, even at the risk of bending it out of shape. This attitude runs so deep it is barely recognisable as an attitude at all. Convincing ourselves that recent developments in computing or genetic engineering are no different from any other kinds of innovations, we acquiesce in the myth of progress that drives the technosciences forward. But this is to misunderstand the effect that technoscience is having and will have – not only on 'the natural world' but also on our humanity.

The danger is not that we create a monster that runs amok, or a plague of zombies, or a rogue AI – or a planet of the apes, for that matter – but that we begin to see ourselves and others as something less than fully human, as machines to be rewired or recalibrated in line with the dominant ideological worldview. In that case, we would *already* have arrived at a perilous situation – a situation where our perception of ourselves as bounded by and connected through nature had given way to the 'post-humanist' view that humans are fleshy automata, subject to endless modification. Denuded of the ancient idea that humans are deserving of dignity by dint of being humans, we would have entered the liminal realm of the uncanny.

'My form is a filthy type of yours,' says Shelley's monster, 'more horrid even from the very resemblance.' *Here Be Monsters* is about what we cannot see and may not recognise until it is too late: the process by which the technosciences are redefining humanity in line with their own mechanistic assumptions, in ways that may prove fatally corrosive to our solidarity and flourishing. It is a book about technoscience and capitalism, but ultimately it is a book about *us* – about the kinds of creatures we are, and are not. It is a call to question our direction of travel, to ask where it might be leading us, and to consider what we might become in the process.

It is about fighting back, through *thinking* back, against the monsters at the edge of the map, in the name of a radical humanism.

PART I

THE TYRANNY OF ABSENCE
Technology and Human Connection

[N]o human life, not even the life of the hermit in nature's wilderness, is possible without a world which directly or indirectly testifies to the presence of other human beings.

HANNAH ARENDT

1

FROM MEATSPACE TO THE METAVERSE

Connection and Disconnection in the Digital Sphere

Mark Zuckerberg – the world's most recognisable tech bro – is standing in a virtual mansion. Around him are empty suits of armour, astronautical kit, bookshelves and a floating fireplace. The mansion affords spectacular views of a tropical paradise (to the left) and snow-capped mountains (to the right). It looks like it might be a Bond villain's lair, though Zuckerberg himself is conspicuously lacking in the evil charisma of a Blofeld or a Goldfinger. In fact, he is lacking in charisma, full stop. Even before he dons his avatar and 'teleports' to a virtual spaceship to play poker in zero gravity (as one does), he seems robotic, a little uncanny even. The co-founder of Facebook is clearly uncomfortable trying to appear comfortable. He cannot quite ... *connect*.

Yet what Zuckerberg lacks in pathos, he more than makes up for in ethos. Whatever you think of him, here is the guy who turned a drunken college experiment into the world's most popular social media platform, changing the world in the process. His advice to his fellow tech entrepreneurs, to 'move fast and break things', is era-defining – an encapsulation of Silicon Valley's revolutionary creativity and arrogance. Perhaps more than any other living person, he has transformed human social life, from the way we respond to politicians to the way we receive (or 'consume') information. But above all he has changed the way we talk. The guy who can't quite connect with us has changed the way we connect with each other.

Today he has big news to share. Facebook, Inc. is changing its name, and changing its business model, too. From now on it will be known as

Meta Platforms, Inc. and its goal will be to develop 'the Metaverse' – a fully immersive internet experience in which we will be able to socialise, work, shop, worship, fall in love, and do all the other things we currently do in the physical world, which is to say the analogue world of physical bodies in physical space. Utilising Meta's own reality-building platforms, VR headsets and augmented-reality glasses, we will cease to think of the internet as something we log onto via a laptop or smartphone and begin to think of it in three dimensions, as an extension of our physical lives. We will even be able to hold digital 'objects'. As one of Zuckerberg's developers explains, the Presence Platform will give the denizens of the Metaverse the feeling of *embodied* experience – the key, he says, to 'feeling connected'. Zuckerberg himself has described this experience as the 'holy grail of social interactions'. We are standing on the cusp of a revolution.

'Our mission remains the same,' says Zuckerberg. 'It's still about bringing people together.' And yet, if the Metaverse becomes a reality, our togetherness will be qualitatively different than it is on social media platforms à la 2023: our togetherness will have *physical* substance, or the sensory illusion of such. Indeed, if the Metaverse becomes a reality, it will *become reality*, or as close to reality as makes no difference.

As Zuckerberg puts it, 'It's gonna be so cool!'

To say that media commentators were unimpressed with Zuckerberg's promotional video would be to put it delicately. As well as noting the Facebook founder's awkward impression of a human being, many journalists raised the possibility that his rebranding had less to do with reimagining the internet than with distancing his company from its reputation as a cutthroat, amoral monopoly, responsible for spreading ethnic hatreds, rightwing populism and poor mental health. Just a few weeks before the video was released, in October 2021, a former data analyst named Frances Haugen had testified before a US Senate committee that Facebook's own internal research had revealed how its algorithms were linked to social ills such as misinformation, conspiracy thinking and political polarisation, and had actively tried to hide what it knew. With the Democrats still gunning for Facebook due to the Cambridge Analytica scandal (where users' data was illegally harvested and used to 'micro-target' voters ahead of the 2016 election), and some Republicans still smarting from Facebook's decision to ban former president Donald Trump from its platforms in the wake of the Capitol riots, a concerted public relations drive was a matter of necessity if Zuckerberg wanted to calm investors' nerves. His Meta pitch looked to some like an expedient move made with one eye on the trading floor, where the personnel are famously sensitive to the 'likes' of the public at large.[1]

That was nothing, though, compared to the rotten tomatoes thrown at Zuckerberg's vision of the Metaverse itself – which, like the supervillain lair from which he delivered his message, managed to be both nerdy and naff. As many technology commentators noted, metaverses of one kind or another are hardly a new idea in tech circles, and Zuckerberg's version was nowhere near as original as one might have expected in a world where 'brain–computer interfaces' are all the rage in futurism. SpaceX founder Elon Musk, who would soon break into the news cycle himself with vision of a digitally enhanced macaque playing classic Pong with its brain alone, was dismissive of the very idea of the Metaverse.[2] Internet activist Ethan Zuckerman suggested that Zuckerberg's iteration was not much more ambitious than the one that he, Zuckerman, had conceived and built in 1994, and certainly little improvement on Second Life, the online multimedia platform that caught the world's attention in the early noughties. To be fair to Zuckerberg, trying to convey a three-dimensional experience in a two-dimensional video is a bit like trying to advertise colour television on a black-and-white set. But even allowing for the technical challenge, the Metaverse graphics were underwhelming. The Metaverse, wrote Zuckerman, 'looks like junk'.[3]

So, not great publicity for Zuckerberg. But despite the mockery and cynicism levelled at the centibillionaire, one point remained unremarked upon, or inaudible amid the hot takes and one-liners. It can, I think, be stated simply: if by the Metaverse we mean those technologies that transfer information to digital spaces, especially social digital spaces, then not only do we *already have one*, but we are also clearly suffering as a consequence of it.

The Annihilation of Space

Named for the expanse of virtual real estate in Neal Stephenson's 1992 novel *Snow Crash*, the metaverse does not describe an 'embodied internet' but a disembodied reality. It is closely related to cyberspace in the sense that the novelist William Gibson used that term in *Neuromancer* (1984), another science-fiction fantasy with a markedly dystopian flavour. In other words, the metaverse is the opposite of 'meatspace' – a coinage often credited to Gibson, who contrasts the world of actual ('meat') bodies to the 'consensual hallucination' of cyberspace. In this sense, and rather pleasingly, 'meta' is an antigram: like 'harmfulness' for 'harmless fun', it's an anagram of its own antonym.

The more common names for *our* metaverse are the internet and the World Wide Web: the networking infrastructure, or network of networks, that allows computers to connect to one another across the world, and the information-sharing system built atop it. Developed in the 1970s and 1980s respectively, these two systems have transformed our world, and their defining trait is not the *presence* of others but the *absence* of others, or of *embodied* others – a fact no amount of 'haptic technology' (technology that uses tactile sensors to reproduce the sensation of touch, as in virtual reality gloves) can gainsay. Yes, they might connect us in many ways, but they also keep us apart. They are technologies based on the circumvention of meatspace.

In this sense, the metaverse is no different from other media and communications technologies. The whole point of such technologies is to make remote connection possible. A printed book takes ideas from one individual or society and reproduces them in another context, while an old-style telephone takes sound waves from one person, turns them into electrical waves and propagates them over an electrical wire, before turning them back into sound. All are designed to contain and convey information to people who are not present to us.

So fundamental are such technologies to the modern human environment – so habituated are we to them – that it is easy to forget just how recently they appeared in the history of *Homo sapiens*. For the great majority of our species' history, and indeed our species' prehistory, communication was necessarily conducted in circumstances of physical presence, initially in a 'conversation of gestures' (displays of aggression, submission and so on), and later in symbolic language, in which sounds come to stand for particular concepts.[4] We spoke face to face, and the content of our messages derived from the symbolic meaning of our words and the variety of physical gestures and expressions that accompanied those utterances: bodily orientation, eye contact, proximity, intonation, tempo, touch. The emergence of visual symbols some 40,000 years ago, first in the form of cave paintings and then in rock carvings, or petroglyphs, allowed for meaningful interactions between people separated in time or space – but even then most communication would have happened person to person. It is only after the emergence of writing, around 5500 years ago, that the world as we know it appeared on the horizon, and not until the invention of printing, a mere 600 years ago, that it began to take recognisable shape. From that point, development was exponential (in what we now call the developed world, at least). In the space of just a few hundred years we went from Gutenberg's converted wine press to the first newspapers in the 1600s; the telegraph, the photograph, the phonograph, the telephone, film and radio in the 1800s; television in the 1930s; the mainframe computer in the 1940s; videotape in the 1950s; microprocessors in the 1970s; the World Wide Web in the 1980s; and social media in the 1990s. And with each stage of this technological development, the way human beings connect with one another, and with the wider society, was transformed.

That last point is enormously important when considering the history of all technologies, but communication technologies especially, because they bear on the distribution of knowledge and the way humans interact. Though not the only factor involved in the transformation of communication (one must be careful to avoid the trap of technological determinism), it is clear that *what we say to each other* depends at least in part for its effect on *the medium through which we say it*. In the 1960s, the Canadian philosopher Marshall McLuhan expressed this idea in his famous maxim, 'the medium is the message'.[5] What he meant was that the meaning derived from any message that comes via a communication technology depends in part on the character of that technology. Thus, a social media post consisting of an image and a pithy caption not only imparts information to one's followers but also helps to establish a context that *itself* transforms society. As the great media theorist Neil Postman (McLuhan's student) was fond of saying, technologies are 'ecological': they change the human element in the same way that a single drop of red dye changes a glass of clear water. Insert the printing press into Europe and you don't get old Europe plus the printing press. You get a different Europe altogether.[6]

Johannes Gutenberg's printing press, invented c. 1440

Let's take the printing press as an example of how forms of communication can change society. Over many years, and in combination with a range of other influences, printing led to the democratisation of religion (for finally God-fearing Europeans could read the Bible in a language they understood); to the spread of literacy to the unwashed masses (most of whom remained illiterate, and unwashed, until well into the twentieth century); and to the emergence of the 'public sphere' – that notional agora, or meeting place, in which philosophical, political and scientific ideas can be set down, shared and argued over by those who are not physically proximate. And from those developments flowed further developments. For example, and as we'll see in the next chapter, one crucial effect of the public sphere was to make possible the 'imagined communities' that the historian Benedict Anderson saw as central to the rise of the nation-state. People who would never meet in person began to identify as members of this or that nation or culture. They became communities of the likeminded, bonded through imagination, and from this a whole new kind of politics – much of it catastrophic for the less 'developed' cultures of the world – became possible and permissible.[7] When commentators point to the downsides of new technologies, they are often accused of spreading 'moral panic', and sometimes the accusation is fair. But an approach to technology that does not allow for any downside at all is far more dangerous, as those on the bayonet end of empire would no doubt testify, if they could.

For Postman, who could panic with the best of them – not least when it came to the effect that television was having on his fellow Americans – an especially significant development in communication technology was the introduction of telegraphy, which led to what the Boston *Patriot*, in one of the first telegraph messages ever sent, described as 'the annihilation of space'. That's to say, the telegraph severed the relationship between transportation and communication, connecting people as never before and putting the Pony Express out to pasture. For the aptly named Postman, this also led to the severance of information and purpose, as the former began to appear indiscriminately, in what he calls an 'information glut'.[8] But the more fundamental point, and certainly the more significant one in terms of where we are today, is that telegraphy and telephony make thinkable a 'social media' whereby people can converse in something approaching real time. And it is this *disembodied* sociality that we, embodied creatures that we are, are finding it very hard indeed – for some deep reasons and some more proximate ones – to navigate.

Addicted to One Another

'Oh this is going to be addictive ...'

Dom Sagolla's first tweet, the thirty-eighth ever sent, could not have been more prescient. The mission statement of Twitter, Inc. is 'to give everyone the power to create and share ideas and information instantly, without barriers' – a vision that would surely appeal to any social media company, keen as they are to cast themselves as the liberators of the public sphere. And yet Sagolla, Twitter's co-creator, thought first of its addictive qualities when taking to the platform himself on 22 March 2006. Twitter was only one day old, but '@dom' had already identified one of its most pernicious qualities.

Even before the launch of Myspace and Facebook in the early 2000s, the idea that public digital spaces might prove addictive was in the mix, as writers and social commentators intuited the compulsive nature of the web. In Stephenson's *Snow Crash*, for example, access to the metaverse is controlled by a media magnate called L. Bob Rife, who attempts to manipulate its denizens through a program that is half-virus and half-drug. Similarly, the protagonist of Gibson's *Neuromancer*, Henry Dorsett Case, longs to plug himself back in to the virtual reality 'dataspace' from which he has been expelled for stealing from his employer. Even the metaphor of *the web* itself reveals a deep ambiguity with respect to digital spaces, implying both freedom and captivity. The orb-weaver spider in my Fremantle backyard seems perfectly happy with her (trampoline-sized!) creation, but the wasps and flies that get tangled up in it are rather more equivocal.

Sagolla's candour notwithstanding, the developers of social networks characterise their users as spiders, not flies. Social media platforms are simply 'tools' that can be used for good or bad ends.[9] Yes, such platforms *can* be addictive, but so can soap operas and pistachio nuts. Indeed, the idea that people can be manipulated is an insult to the professed ethos of the Silicon Valley entrepreneur. People have a choice, after all, about where to spend their time and money, and it's arrogant and paternalistic, not to say ruinous to the bottom line, to go around declaring otherwise.

Such libertarian pleas are nonsense, of course. Due to a spate of mid-profile defections from Silicon Valley in recent years, we now have plenty of information on how social media platforms like Facebook utilise the dynamics of addiction to increase their profit margins. In essence, it is an 'attention extraction' model whereby vast amounts of data are collected in order to build profiles of individual users, whose behaviour can then be predicted *and altered* with a view to attracting advertisers.[10] In the words of the author Shoshana Zuboff, it is a market in 'human futures' – a new form of advertising sold on the understanding that predictions can be made on the basis of surveillance.[11] One crucial aspect of this advertising model is that it is not our *stated* preferences but our *revealed* preferences that are so surveilled – not the things we say we like, but the things our behaviour says we like.[12] I may assure you that I only ever use YouTube to watch old lectures and listen to political podcasts. But YouTube also knows my weakness for boxing highlights, old comedy sketches and the PronunciationManual channel. However high-minded my search, it can always tempt me with a two-minute video of Mike Tyson's greatest knockouts, Monty Python's Cheese Shop sketch, or a funny pronunciation of 'synecdoche', with a view to getting more advertising in front of my illuminated face. Nor is this advertising model peripheral to the business of navigating the internet. As many defectors frame it, the model *is* the internet. Even the results from a Google search are tailored to individual users. Information itself is now bespoke.

The best argument against the libertarians who downplay the addictive nature of the internet is the experience of individual users. As every Facebook aficionado or Instagram habitué is aware, social media *is* addictive in the sense defined in most reliable dictionaries, as *the condition of being unable to stop using or doing something as a habit.* (Small coincidence that both social media companies and drug dealers describe their clients as 'users'.)

The notion that behaviour can be addictive was controversial until relatively recently, dismissed by psychiatry as too nebulous to be clinically useful. The conceptual framework was physical: addiction was understood as the interaction of a particular brain (with a particular chemistry) and a particular substance. But today it would be truer to say that behavioural addiction frames the way the psychological professions think about addiction in general. Thanks in part to studies done on soldiers returning from the Vietnam War, we now know that many drug addicts, for example, are addicted less to substances than they are to the circumstances of their taking. It follows that addictions will be harder to beat when the circumstances are harder to escape, and to that extent the ubiquity of the internet is both a cause and a consequence of internet addiction.[13]

There are now countless studies that attest to the deleterious effects of social media on users' mental health, as the emotional stresses of social comparison, self-presentation and low self-esteem combine to engender anxiety and depression, especially among the young.[14] In one of the most sobering interviews in the 2020 documentary *The Social Dilemma*, the social psychologist Jonathan Haidt presents data that show how the rates of suicide and self-harm in pre-teen and teenage girls have risen sharply since 2009, the year that social media platforms became available on smartphones. For teenage girls, the rate of non-fatal self-harm rose by 62 per cent, while for pre-teens it rose by 189 per cent. Since the 2010s, the suicide rates of these groups have risen 70 per cent and 151 per cent respectively. Academics are rightly cautious about confusing correlation and causality, but the evidence is mounting fast that platform capitalism is a crack-house economy in which the first bag *and all subsequent bags* are free, and the users – *users* – are looking pretty poorly.[15]

The question is not whether social media is addictive, or whether it is causing social and psychological problems, but why and in what ways. Calling it addictive is not enough, for it isn't the platforms to which individual users are in thrall, any more than heroin users are in thrall to the drug's taste or powdery texture. They are addicted to the feelings afforded by the stimulus, and the feelings afforded by social media relate to our need for connection with others, for approval, gossip, friendship and love – in other words, to our social being. Why is social media addictive? The clue, it turns out, is in the question.

In his excellent book *The Twittering Machine*, Richard Seymour makes the astute observation that when talking about social media, the purely medical model of addiction is about as meaningful as a medical model of love. Of course, there is a chemical aspect to addiction. We know that when we engage in certain behaviours our brains are 'rewarded' with dopamine. But studying that process in isolation from the behaviour and its social significance is a bit like trying to understand a painting by studying the relationship between the colour- and shape-responsive regions of the brain – fascinating, certainly, but limited in terms of criticism and aesthetics.[16] There is profound disagreement among academics about precisely what addiction is, and I won't venture a definition beyond my superficial one. But I don't imagine for a second that a decisive answer will be found on a brain scan, any more than a meaningful take on *Guernica* will be found on a Taubmans colour chart.

Like many writers in this space, Seymour is interested in the feedback tools social media platforms use to maximise engagement. In particular, he is interested in the various 'like' buttons that allow users to convey approval or disapproval in the (binary) form of a plus or a minus, a thumbs up or a thumbs down. As he notes, the effect of these buttons has been profound. When Facebook's thumbs-up icon was introduced in 2009, for example, user engagement rose sharply. This led to comparisons with the operant conditioning chambers the psychologist B. F. Skinner invented to study animal behaviour.[17] Like the rats and pigeons in Skinner's experiments – or like the Pong-playing macaque in Elon Musk's demo – Facebook's users were being conditioned through positive reinforcement. But whereas the former were conditioned with food (generic pellets for the rats and pigeons, a banana smoothie for the macaque), social media users were conditioned with attention. And, of course, with *in*attention, for as Skinner discovered, a pigeon deprived of its expected reward will not lose interest in the feedback button but tap away at it incessantly, like a cheesy Boomer bobbing his head in time to the bassline of Pink Floyd's 'Money'. And so it proves on social media: in the attention economy of platform capitalism, the feedback button is 'cyber-crack'.[18]

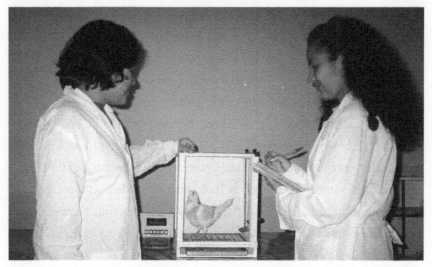

A conditioning experiment on pigeons using a handcrafted operant conditioning chamber, or 'Skinner Box'

Again, it is less the crack than the feelings afforded by the crack that keep us coming back to the platforms, and in this respect the Skinner Box model doesn't take us very far in explaining the social media phenomenon. For while Skinner's rats and pigeons want food, social media users appear to want approval – a very different quantity. One could almost say that they want *one another* – that the real 'substance' the user craves is the company and esteem of other human beings. They simply want to *be* with one another – to overcome the annihilation of space.

This is the deep story of social media. Yes, the 'social industry' (as Seymour calls it) is a world that rewards the performative and even narcissistic aspects of our character. And yes, it turns all of us into micro celebrities compelled to curate our lives and identities for the delectation of our followers.[19] But it also reminds us that without the social there *is no human life or identity* – that the social is always and everywhere *prior* to the emergence of our sense of self. The social media entrepreneurs insist they are connecting us, and it is easy to scoff at their idealism. But in one respect they are on the money. The 'social network' is only effective because human beings are social animals.

A Morbid Sociality

When Margaret Thatcher made her infamous comment about there being 'no such thing as society', only 'individual men and women', she was either being mischievous or stupid. Quite apart from anything else, the idea that society and the individual can be neatly separated is nonsensical. Long before we become individuals with a sense of personal autonomy, we exist in physical community with others – with our parents, most significantly – and it is only through their ministrations that we are able to become individuals at all, in the sense of developing personhood. 'No man is an island entire of itself,' wrote the poet John Donne in his *Meditations*, and anthropologists would rush to agree. The categories of individual and society can no more be separated than light and heat.

Nevertheless, the rise of liberalism in the seventeenth and eighteenth centuries did bring the idea of the individual as a separable *political* entity to the centre of European life. It did so for some excellent reasons. Up until that stage in history, people had basically done what they were told. Everything existed in a 'chain of being', with God at the top and the beasts at the bottom, and the rest of us somewhere in between – and heaven help the scruffy peasant who tried to climb above his station or stray beyond her nursery. Liberalism, however, challenged this worldview. Incubated in the Protestant Reformation (which in the early sixteenth century rejected the universal authority of the Church) and systematised by the philosopher John Locke in the second half of the seventeenth century, it stressed freedom of religion, equality under the law and parliamentary sovereignty. But it was its conception of the individual as possessing certain legal rights and the capacity for autonomous thought and action that really revolutionised society, marbled as it was into an economic system – capitalism – that is based on private property and individual enterprise. Indeed, this conception is now so deeply rooted that it is almost impossible to consider objectively. As far as our way of life and our politics go, it surrounds everything, like radiation from the Big Bang.

So, that's *liberal* individualism, and it clearly had some benefits, though to say that the benefits have been unevenly shared – between women and men, the rich and the poor, the West and the rest – would be an understatement. But in recent decades we have witnessed the emergence of a harsher form of individualism, which has manured the ground for the emergence of Silicon Valley–style thinking about the individual and society, and the relationship between the two. It's the individualism Thatcher was channelling when she made her comment about society not existing. We might call this *neo*liberal individualism.

Political commentators are sometimes reluctant to grant the existence of neoliberalism at all, and it's true the term is often lazily deployed by people who are (quite reasonably) dismayed by the greed, selfishness and inequality associated with capitalism in general. But neoliberalism *does* exist, and it's a very different beast from the kind of capitalism envisaged by Adam Smith in his 1776 treatise *The Wealth of Nations*, the founding document of bourgeois economics. For what neoliberalism does, quite deliberately, is take the ethos of competition that characterises *economic* life and pipe it into every area of society that its priesthood can identify as in need of greater 'efficiency'. Whole areas of life that once existed substantially outside of the market – public services such as healthcare and education, as well as the arts and civil society – are thus subjected to market logic, through metrics, league tables, financial accounting and key performance indicators. Often channelling the 'Social Darwinism' that seeks to draw a comparison between the competition for resources in nature and competition in the economic sphere (a complete misreading of Darwin, by the way, who understood that evolution was as likely to select for altruism as it was for aggressive competitiveness), neoliberalism elevates market logic to the status of a law of nature. It introduces into the evolutionary record a new type of human: *Homo economicus*.

The most profound effect of the so-called 'neoliberal turn' is felt not at the level of institutions, but at the level of humanity itself. As market logic comes to dominate more and more spheres of human life, establishing fixed measurements and indices of how merit and achievement are to be judged, and obliterating values and practices associated with alternative ways of living, something profound begins to occur in humans' sense of themselves. Lacking the grounded relationships and rituals of more traditional societies, and the 'ascribed identities' that emerge from them, they become the 'entrepreneurs of the self' described by the philosopher Michel Foucault in his lectures on neoliberalism.[20] For perhaps the first time in human history, people are obliged to *create* their identities, upon pain of being swamped by the chaos of modern life, with its imperatives to distinguish ourselves, to be *different*. Identity is transformed from a *given* into a *task*.[21]

It is here, of course, that social media really begins to work its dark magic. For identities, like the individuals who possess them, cannot be said to exist at all unless they are validated by other individuals, and it is on social media platforms that such validation now takes place for many. There is something deeply tragic about the way that neoliberal capitalism, having ungrounded humans from the emotional 'infrastructure' of material social life, now erects atop the wreckage a network of anxious and addicted denizens who are incessantly mined for their (monetisable) data.[22] The addictive relationship stands in for the relationships that have been undermined by capitalism, but is also the principal means through which capitalism spreads, further undermining social life as it goes.

At the same time that neoliberalism was becoming the only game in town – propagated not just by Thatcher and Ronald Reagan but also by the later centre-left governments of Paul Keating, Tony Blair and Bill Clinton – the personal computer became a feature of the 'knowledge economy' that was supplanting the industrial one. Where capitalism had once required workers to congregate in noisy, smelly factories, it now required them to sit in offices and deal with data. Slowly but inexorably, relations of absence came to replace relations of presence in many developed economies, as industry was off-shored to places where the workers were cheaper and kicked up less of a fuss. A new weightlessness came to characterise the economy, as the neoliberal principles of financialisation and globalisation – themselves made possible by innovations in algorithmic technologies – helped capitalism break the bonds of post-war social democracy. The 'real economy', as it's sometimes called, was replaced with what felt like a virtual one.

So there is a symbiotic relationship between neoliberalism and technologies of absence. But there is also a sense in which those technologies undermine the idea that remains on the cover of tech-capitalism's brochure: that we are all free agents in the great democracy of the internet – that we are the spiders on the web, not the flies in it. On this logic, computers are not merely extensions of humans' intentional actions, but a model for what we *are*.

And it is this model of the human animal, not the cosy liberal one, that really underpins the metaverse.

The Human Machine

Neoliberalism has always considered humans and human societies as computational. For while neoliberals will often speak about 'choice' and 'economic freedom', conventional neoliberalism conceives of society as little more than a gigantic calculating machine. For the academic and author Will Davies, this can be seen in the philosophy's founding documents, in particular the work of Austrian compatriots Ludwig von Mises and Friedrich von Hayek, who in the 1930s and 1940s represented the capitalist market 'as a type of man-made technology for the performance of calculation'.[23] Humans are reduced to calculating entities, instinctively weighing costs against benefits in the great bazaar that is human society. Game theory and rational choice theory, which both provide mathematical frameworks for analysing human interactions, are extrapolations of this strange worldview, and, as we'll discover later, are related to the development of nuclear technologies and the need to evolve military strategies that could keep the Cold War from turning Hot. But for now let us only note that at both the micro and the macro level, neoliberalism is a deeply inhuman creed – an algorithmic creed, no less. In the words of Paul Mason in *Clear Bright Future*, 'submission to the logic of the market becomes a gateway for submission to the logic of the machine'.[24]

In the 1980s and 1990s, as computer technology moved to the centre of economic and cultural life, academics and developers in a number of disciplines began to make explicit connections between that technology and the human brain. An idea began to take hold: humans were not the free and rational agents of liberal and neoliberal legend but complex systems responding to stimuli. In neuroscience, computer science and cognitive psychology, it became fashionable to assert that the human mind was 'just' a computer, albeit a fantastically complex one.[25] Today, this idea has great currency. For the bestselling historian and intellectual Yuval Noah Harari, for example, there is little essential difference between a machine that makes a cup of tea and the person who, by pressing buttons, sets the tea-making process in train. Both are computers, albeit fashioned from radically different stuff.[26]

A cursory glance at the pronouncements of influential entrepreneurs confirms that this view is immensely popular in Silicon Valley.[27] The reason isn't hard to fathom. If humans are considered sophisticated computers, the story of humanity becomes one of increasing complexity, and technology becomes the inevitable manifestation of human advancement. This view not only flatters the avatars of technoscientific capitalism, but also sanctions their reprehensible conduct. For if the human brain is a souped-up iMac, then it must be possible, and even permissible, to hack it.

And of course that's what they've done, through platforms that embed social life within an advertising model. This model auctions off our attention to the highest bidder and also *demands* that attention – not in the way of the old-style telephone's piercing tantrums in the empty hallway, but constantly, inexorably, and with a view to extracting yet more profit. The essential nihilism of capitalism – a system that reduces every object to the status of a commodity, and in so doing empties it of any 'sacred' content it might once have possessed – is ushered to the very heart of what makes us human: our ability to talk to one another, to share our thoughts, to listen and to love.

But this is not the end of the story. While humans do respond to some stimuli in a way that superficially resembles Dr Skinner's pigeons, or

Musk's banana-crazed macaque, *we are not reducible to those responses.* No, we are far more interesting creatures than the entrepreneurs of platform capitalism can imagine, with their boringly one-dimensional vision of a three-dimensional metaverse, and a human community denuded of presence – of the messy materiality of existence. Neither calculating nodes in the neoliberal marketplace nor maddened rats in the conditioning chamber, we are social, creative, intentional animals whose reasoning ability and pre-rational desires exist in a precarious balance.

In cleaving to the view that humans are complex machines, the black-skivvied bros of Silicon Valley not only justify themselves to themselves but also perpetuate a view of humanity that is corrosive of human flourishing. As we'll see, such a view is part of a new and dangerous dualism that regards human minds as separable from the bodies in which they are housed. The view is absurd, but it is widely held, and there is nothing to say that we will not try, through medical and other interventions, to transform ourselves into our own avatars.

'There are only minds,' murmured a TV commercial for an internet network in 1997, the same year that the first social media platform, SixDegrees.com, came into being. It was wrong, but it was also prescient. For in the following decades social media companies did construct a network of minds, idealistically and even naively at first, and then with a view to monetising the data gleaned from the billions of messages sent in the course of any day. Building on the fundamental desire for meaningful connection with others, platform capitalism thus proceeded to remake it through its circuits of surveillance and control. And the result was a *morbid* sociality, a *monstrous* sociality even – a sociality that can never become the thing it tries to counterfeit.

No wonder the guy in the virtual mansion finds it difficult to connect. Welcome to the Metaverse. *So cool!*

2

SOCRATES IN CYBERSPACE

Technology and the Public Square

Perth, Western Australia: 7:13 a.m. US Tweeters are still chirping. A Republican senator has described herself as a Christian and a nationalist, drawing much talk of Nazis from progressives, and prompting one of the yammering busts from Fox News to say something moronic in her defence. #idiocracy is trending.

8.37 a.m. Stung by recent criticism of his forays into commentary (something about 'cultural Marxism' and leftwing bias in political media coverage), an Australian reporter has stirred up Twitter by describing its progressive users as 'cockroaches'. This is the cause of widespread hilarity.

10.01 a.m. I read an online article on the topic of transgender activism. Hundreds of comments unspool beneath it. Some are helpful clarifications, or requests for such, in the spirit of the piece, which is challenging but neither bigoted nor nasty. Others are angry. Some are abusive. One accuses both the author and the website of placing trans kids at greater risk of suicide.

12.22 p.m. On Twitter, some users are incorporating 'cockroach' (now trending) into their usernames. The cockroach icon has become a badge of pride: progressive Twitter is infested with them.

1.42 p.m. The transgender article has now hit Twitter. One Tweeter calls its author a 'c**t' and wonders if he was dropped from a balcony as a child. There are many more posts in a similar vein. An author I like, and who should really know better, has attacked the piece without reading it, receiving many likes from his followers as a result. I compose a response and then delete it.

3.55 p.m. #drongo is trending in Australia. I resist the temptation to follow the link.

3.57 p.m. I click on #drongo. Fair enough.

5.01 p.m. The UK is waking up to the news that a politician has been stabbed in his surgery by a young man 'radicalised on social media'. A rightwing pundit has blamed the left.

6.05 p.m. #dickhead is trending. Fair enough.

And on it rolls, and on and on ... As the day progresses, the targets change and controversies come and go, but everything is suffused with the same weird mix of intimacy and thuggishness. It is a city that never sleeps, like Las Vegas: a place of lurid lights and compulsions, of countless small losses of self-control, where the casino always comes out on top. Virtue signalling and groupthink abound. To the left, the right is vile and stupid. To the right, the left is smug and herd-like. True, there is more agreement than not, but in the quantum realm that is social media, the echo chamber and the spotty troll are entangled aspects of the same derangement. Yes, I know how this must sound, and that the 'republic of letters' is full of writers who have clearly mistaken the state of the world for the state of their own bank balances and/or livers. But honestly, what is this crap? In its effect on politics and ideas, social media comes about as close to the ideal of free and reasoned debate as an episode of *Love Island* comes to *Romeo and Juliet*.

It wasn't supposed to be like this. In the late 2000s and early 2010s, social media platforms were considered by many to be the Next Big Thing in democracy – a phenomenon that would reinvigorate and even reinvent the public sphere. Books such as Yochai Benkler's *The Wealth of Networks* (2006) and Clay Shirky's *Here Comes Everybody* (2008) celebrated the transformative potential of peer-to-peer networks and the digital commons, arguing that cheaper social technologies would change the way social networks functioned by allowing groups of likeminded people to bypass traditional institutional models. Certainly they could cite plenty of examples of such self-coordinated action, from image-sharing sites such as Flickr to 'collaborative production' projects such as Wikipedia. But for many readers it was the political cases that highlighted the radical potential of the technology – the flash mobs of Belarus, for example, where protesters opposed to the Lukashenko regime would suddenly appear in public spaces and engage in some innocuous activity such as eating ice-cream, or simply smiling. As the police moved in on the grinning idealists, the thought began to occur to some that perhaps the new tech was implicitly threatening because it was implicitly democratic.

It was the protests that erupted across North Africa and the Middle East in 2011 that catalysed this democratic potential. Prefigured by the 'Twitter Revolution' in Iran, the Arab Spring was fomented in the virtual spaces of Facebook and Twitter before spilling out into the public squares of Egypt, Libya, Syria and Yemen. In Tunisia, where the protests started, it was not inspiring oratory or skilful political organisation that brought the population onto the streets, but images of a street vendor who'd set himself on fire as a protest against corruption and police harassment, while in Egypt a Facebook memorial page dedicated to a young entrepreneur beaten to death by police became the virtual rallying point. Just as fax machines and photocopiers had hastened the end of the Soviet Union – allowing enemies of the communist system to share intelligence and dissident literature – so the algorithms of social media catalysed the Arab Spring. In the words of one Egyptian protester, 'We use Facebook to schedule the protests, Twitter to coordinate, and YouTube to tell the world.'

There was more than a hint of technological utopianism surrounding these momentous events, stemming from the apparent convergence of technological means and political ends. While previous democratic springs – from the Springtime of the Peoples in 1848 to the Prague Spring of 1968 – relied on top-down organisation, the uprisings of the 2010s were largely hierarchically flat, self-coordinating affairs. In the movements that followed the Arab Spring, this characteristic became an explicit principle. In the Occupy Movement, for example, the very idea of leadership was rejected, and even concrete demands were frowned upon, on the basis that a specific program would legitimise the very system of power and hierarchy the movement sought to challenge. The slogan 'We are the 99%' referred to the economic injustice that concentrated half of the world's wealth in the hands of just a few individuals. But it was also the expression of a new connectedness – not a class or an identity, but a sprawling *network*.

'Bliss was it in that dawn to be alive,' wrote William Wordsworth of the French Revolution. Did the half-Syrian Steve Jobs feel a similar pang, as he watched events unfold from his sickbed? It would be nice to think so, if a little maudlin. At least it is some comfort to know that he wasn't around for what happened next.

How did we get from the 'Twitter Revolution' to the toxic ecology of contemporary online politics? In asking this question, I am not suggesting that social media is all to blame for the frenzied and superficial character of the online public sphere, still less that there is some ideal example of democratic deliberation (historical or imaginary) to which we should all aspire. The public sphere is subject to all kinds of pressures – some of them obvious, like media bias, and some of them deeply structural, like the existence of particular classes whose members do particular kinds of work that influence how they see the world. All of these phenomena shape and colour the discursive mix. My point is that communication technologies are also marbled into that mix, and that in the journey from Cairo's Tahrir Square to @MarxistCockroach and #drongo, something has happened to the public sphere that is inseparable from the algorithms in which it is now (substantially) embedded. Few readers would disagree with Andrea Nagle's observation (made in her book on the online culture wars and the rise of the alt-right, *Kill All Normies*) that there is 'something about' social media that is conducive to 'the vanity of morally righteous politics and the irresistible draw of the culture wars', or with Richard Seymour's suggestion that there is 'something about' social media that 'magnifies our mobbishness, our demand for conformity, our sadism, our crankish preoccupation with being right on all subjects'.[1]

The question is: what is that 'something' and what is it 'about'?

It's a question of immense importance. In the last chapter, I argued that social media is the cause of a morbid sociality. In order to confront that problem, we need an intellectual culture that is both democratic and productive – a vigorous, combative public sphere in which the vigour and combativeness are not achieved at the cost of sharp analysis. Unfortunately, politics and the public sphere are not themselves immune from the morbidity that characterises online sociality, and so we need to identify the ways the algorithms are fuelling the outrage, prejudice and self-absorption that plague the political cybersphere. That is not to call for consensus or bipartisanship or any of the other niceties favoured in liberal-democratic press galleries, with their technocratic view of politics. I *like* disagreement, and I think it's important. But the fact is that a public sphere that derives its energy from the empty calories of spite and likes and look-at-me zingers is no good for our intellectual health. As Goya put it in one of his satirical aquatints, 'the sleep of reason produces monsters'.

We'll get to the monsters soon enough. But I want to begin in another time and place, with a man whose intellectual health was rarely in question, and who had the distinction (if one can call it that) of only being ugly on the outside.

#donkeylips

Socrates wasn't known for his good looks. According to many of his contemporaries, he looked more like a satyr than a human being, with bulging eyes, an upturned nose, flaring nostrils and lips like a donkey's. What he lacked in physical beauty, however (and in bouquet: he wasn't a bather), he more than made up for in cerebral elegance and oratorical ability. As one of his followers, Alcibiades, put it, those donkey-like lips 'had power to entrance mankind'.[2]

Today we tend to associate philosophy with written language and assume a deep relationship between the paraphernalia of textual composition (the numbered paragraph, the aphorism) and careful, purposeful contemplation.

49

Søren Kierkegaard, Friedrich Nietzsche and Ludwig Wittgenstein were all great writers, as well as great philosophers. (Hegel, by contrast, couldn't write for sour apples.) But Socrates' genius was not tied to his ability with pen and paper, or stylus and papyrus. In fact, he didn't hold with *writing* at all.

In the heady days of the 2010s, the tech utopians would often have fun with this fact. Socrates, it was said, was representative of the folly and elitism of technological scepticism – the patron saint of stuck-up intellectuals who panic whenever new technologies threaten to extend ideas to more people than was previously the case. Reasonably enough, they pointed to the irony that we wouldn't even be aware of Socrates' thoughts on writing were it not for the fact that two of his students – Plato and Xenophon – had the presence of mind to commit them to their wax tablets. (Clay Shirky favoured the example of Johannes Trithemius, aka the Abbot of Sponheim, who in 1492 wrote a passionate treatise defending and exalting the scribal tradition. It was printed in 1494.) To put it in McLuhanesque terms, the medium, they argued, undermined the message.

But Socrates was asking the right question. He was looking carefully at a technology and asking, *What does this technology add to our lives and what does it take away from them?*

The key conversation is with the aristocrat Phaedrus, as related by Plato in his *Dialogues*. Referring to another conversation – between an ancient Egyptian king called Thamus and the ibis-headed god Theuth (Thoth) – Socrates argues against Theuth's claim that writing is an aid to better memory and a cure for faulty reasoning, agreeing with Thamus that it has the opposite effect. For one thing, people's memories atrophy, as they come to rely on 'the external aid of foreign symbols, and not the internal use of their own faculties'. Socrates then turns to the question of wisdom:

> For this, I conceive, Phaedrus, is the evil of writing, and herein it
> closely resembles painting. The creatures of the latter art stand before

you as if they were alive, but if you ask them a question, they look very solemn, and say not a word. And so it is with written discourses. You could fancy they speak as though they were possessed of sense, but if you wish to understand something they say, and question them about it, you find them ever repeating but one and the self-same story. Moreover, every discourse, once written, is tossed about from hand to hand, equally among those who understand it, and those for whom it is in nowise fitted; and it does not know to whom it ought, and to whom it ought not, to speak. And when misunderstood and unjustly attacked, it always needs its father to help it; for, unaided, it can neither retaliate, nor defend itself.[3]

It is Socrates' personification of writing (and of painting) that is so clever here, in that it draws our attention to the very thing that written language cannot do: *hold a conversation with us*. And for Socrates, it is through conversation that genuine wisdom is attained. Distinguishing between the content and the form of media, he is thinking about the relationship between communication technologies and culture. By contrast, the tech utopians make the mistake of treating such technologies as more or less efficient means of informational and intellectual transmission. The more transmission the better, say the tech-bros. Not necessarily, says the gadfly of Athens, whose view, while not deterministic, is at least alive to the possibility that the form of the technology will affect how we think.

Was Socrates right about writing? In point of memory, he surely was: as Lynne Kelly shows in *Memory Craft* (2019), cultures that lack written language have fantastically effective methods of committing information to memory, and better memories as a consequence.[4] In point of wisdom, probably not: I think we gain more from writing than we lose, notwithstanding that we *do* lose something. But whether he was right or not, the point is that he was asking the right question: what is given, and what is taken away?

Importantly, he was asking it in the public square – in what the ancient Athenians called 'the agora'. The agora was a meeting place and a marketplace: a *physical* space in which speeches were given, ideas tested and students instructed. Socrates' rejection of writing was founded on the value he placed on intimate contact with his students, and the method of enquiry he gave us – a method stressing scrutiny and questioning rather than authority – emerged from precisely that intimacy. As a technology affording remote communication, written language cuts across that intimacy in ways that affect the *character* of interaction. Socrates' 'dialogic' technique remains embedded in Western academia – in the advice to students to strengthen their arguments by including alternative points of view, for example – but it is also clear that writing, printing, television, the internet and so on not only bring the *virtual* agora – the online public sphere – into being but also affect its qualities.

Detail from Raphael's fresco *The School of Athens* (c. 1508–11).
Socrates is standing sixth from the left.

The point is that communication technologies colour the way we think about the world in exactly the way Neil Postman suggested with his analogy of a drop of red dye introduced into a glass of water. Just think of the way television has transformed politics and the public sphere. Since TV is a visual medium, the appearance and demeanour of politicians comes to take on greater importance, even to the detriment of persuasive argument. Often cited in this connection is the presidential debate between John F. Kennedy and Richard Nixon in 1960 – a debate in which Nixon was generally regarded as having got the better of Kennedy, but lost on account of his sweaty demeanour. For Postman himself, who wrote about television at length in *Amusing Ourselves to Death* (1985), the rise of the image is in this sense linked to the decline of rational argument. Channelling McLuhan (Socrates to his Plato), he argues that particular mediums can only sustain a particular level of ideas, and that television drives reportage towards the abattoir marked 'entertainment'. Consequently, it also changes the way the consumer of news and commentary relates to the world beyond the screen. Postman observes how televised news presents a series of 'and now this' moments, as photogenic newscasters ('talking hairdos') pivot from one issue to another. (Sometimes they pivot literally, turning in their swivel chairs to face another studio camera.) Postman invites us to see such moments as the signature of televised news in general, focusing on one issue after another in a way that leaves all under-analysed. Of course, one could claim that whatever the drawbacks of televised (or simply *visualised*) news, we are still better informed than we were in the past, simply because of the volume of news. But that would be to miss the point that the constant flow of information undermines the relationship between human action and important events, radically changing what Postman calls 'the information– action ratio' – that is, the structure of human responsiveness that determines how much and what sort of information is usefully assimilable. Put simply, it turns us into *spectators*: passive consumers of information, rather than participants in the creation of knowledge.

The passivity that emerges from a public sphere dominated by television is for Postman likely to trigger feelings of hopelessness and powerlessness in the viewer, and no doubt those feelings are partly to blame for the nastiness and superficiality of politics in the cybersphere. But as an interactive medium, social media is very different from TV, and its effects on the public sphere are even more profound. I'll come to those effects later on, but for now let us pause to reflect on the distance we have travelled since Socrates wandered through the agora.

Here, the ugliness is of a different order.

Political Monsters

Friday, 20 January 2017: Inauguration Day in the United States. The crowd stretching out from the west front of the Capitol: the great and the good and the not-so-good rubbing shoulders on the balustrade. The men in suits and long dark coats. Hillary in white. Michelle in red. Melania in powder blue. Former presidents doing their best to look cheerful. An aw-shucks look from Bill. A stiff little wave from Dubya.

And, down the front, the man himself: hair like a sample of loft insulation, a small orange hand on Lincoln's bible, repeating the words once spoken by George Washington: '... and will to the best of my ability, preserve, protect and defend the Constitution ...'

Donald John Trump. The billionaire. Former reality TV host. Filer of bankruptcies. Grabber of pussies. And, now, the 45th President of the United States – its Commander in Chief.

It would be absurd to claim that social media was solely to blame for the election of Trump. The Cheeto Jesus was carried to power by a range of interlocking factors, including the global financial crisis, the failures of neoliberalism to provide economic security, the technocratic character of modern politics, racism, misogyny and xenophobia. But an analysis that excluded the internet and social media would be a partial one. If all tyrannies begin as 'festivals of the depressed', digital media is a Rio de Janeiro of toxic ideological potential.[5]

How are the internet and social media implicated in the rise of populism, and in the broader political polarisation of which it is one manifestation? One answer to that question is simple enough, and stresses the very characteristic the tech utopians took as a positive: the democratisation of information. According to this explanation, the internet has created a situation in which quantity and quality are dangerously out of alignment, as what Postman called the 'information glut' that began in the nineteenth century becomes 'information meaninglessness'. Add in the fact that the people receiving the information may lack the knowledge or cognitive ability to separate good arguments from bad ones, and the crisis becomes acute. As one (very grumpy) commentator put it, 'people who have no idea how to make a logical argument cannot realise when they're failing to make a logical argument'.[6]

In the months after Trump was elected, this style of explanation was much in evidence, as some commentators speculated on whether Trump's oratorical style – his 'tortured syntax, mid-thought changes of subject, and apparent trouble formulating complete sentences, let alone a coherent paragraph, in unscripted speech' – was due to a decline in his cognitive ability or a tactic designed to woo his base.[7] Others opined that Trump's supporters were more vulnerable to 'biological instincts' such as tribalism due to their lack of formal education, while still others invoked the 'Dunning–Kruger effect', which purports to explain the illusory superiority experienced by people of low cognitive ability.[8] In this way, the internet gave a new lease of life to an old idea – the idea that the demos (voters) cannot be trusted with democracy. The idea is most famously associated with Plato, who in *The Republic* argued that democracy leads to rule by the few or rule by the mob, and that a ruling caste of philosopher kings is much to be preferred. The commentators holding forth on the stupidity of Trump and his supporters didn't quite go that far, but there was a note of knowledge-class arrogance in their pronouncements.

The 'death of expertise' explanation for Trump's win missed some important points. Yes, low educational attainment was a reliable predictor of support for Trump. But as the statistician Nate Silver suggested, education levels were to some degree a proxy for wider cultural shifts related to the greater esteem attached to education and cognitive skills in an increasingly knowledge-focused economy.[9] Where once it was those workers with practical knowledge and physical dexterity that dominated the economic imaginary, today it is those who work with information that stand (or rather sit) at the centre of production. Many working-class people now feel culturally and economically stranded. Their identities have been relegated.

It is here, in these feelings of wounded status, that social media is such an important factor, not only on the right but across the political spectrum. As neither speech nor writing, but a sort of hybrid between the two, social media combines the personal and the political in ways that are rarely conducive to productive debate. It is no coincidence that one of Trump's favourite words on Twitter is 'unfair': in the febrile world of cyberpolitics, taking things personally is par for the course.

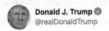

> **Donald J. Trump** ✔
> @realDonaldTrump
>
> So disgusting to watch Twitter's so-called "Trending", where sooo many trends are about me, and never a good one. They look for anything they can find, make it as bad as possible, and blow it up, trying to make it trend. Really ridiculous, illegal, and, of course, very unfair!
>
> 4:41 PM · Jul 27, 2020 · Twitter for iPhone

Tweet from Donald Trump deploring his 'unfair' treatment

Social media transforms the relationship between identity and belief. It was only in the eighteenth century that the modern public sphere became possible: emerging from networks of private letter-writing, it was catalysed by the printing revolution, which established a space of intellectual exchange between strangers. But while this public sphere was largely impersonal in nature, the social media revolution has embedded public discourse in a personal frame. According to the sociologist John Thompson, the result is an intellectual culture that combines features of 'dialogic' interaction, such as conversation and letter-writing, and 'non-dialogic' interaction of the kind found in the public sphere before the internet. Many of the problems associated with social media derive from this hybridity, since it fuses our sense of personhood and our political and moral beliefs. Caught between personal and public registers, we succumb to what Thompson identifies as a new and distinctive form of fragility.[10]

For the political theorist Langdon Winner, it is this confusion of the personal and the political that helps to explain the febrile character of politics in the cybersphere. Drawing on Hannah Arendt's attempt, in both *The Human Condition* and *On Revolution*, to determine what kinds of activity constitute genuine involvement in politics, he distinguishes between the private and the public 'realm'. While the first revolves around intimate relationships and highly personal activities that are self-defined and self-determined, the second concerns politics and society as a whole. It follows that private life is a space in which a person is able to establish and maintain a clear sense of who they are, and that public life is a 'space of appearances' in which identity is revealed through speech and action, by 'what one says and does in public gatherings'.[11] In other words, in the public realm a person's identity is constantly recognised and evaluated by other people.

'From the theoretical standpoint Arendt offers,' writes Winner, 'it is clear that the all-too-common attempts by people in online political communication to define, protect, cling to, and incessantly broadcast their personal identities stands at the root of the various discourse pathologies commonly encountered on the Internet today – "trolling," "flaming," "bullying," and the like.' For Winner, online politics is built upon a confusion of the private and public realms, or the intrusion of one realm into the other. Politics has become an expression of identity in which commitment is subordinate to self-expression, where the self, as a newly fragile entity, is always on the lookout for the social glue that can bind it together for another day, and always on the prowl for the social solvents that would weaken its (shaky) composition. It is not that the personal has become political, or even that the political has become personal; it is that there is now no clear distinction between the two.

To readers on the progressive side of politics, these observations will be grimly familiar. In *Kill All Normies*, Andrea Nagle describes how the 'call-out' culture of online progressives – the obsession with shaming and privilege-checking that characterises left-wing Twitter – has resulted in an atmosphere of mutual surveillance, acrimony and recrimination.

This 'vampire's castle' of 'witch-hunting moralism' (as the late blogger and author Mark Fisher described it) derives in part from the cultural turn in left-wing politics in the 1970s, when racial, sexual and gender identity began to replace economic class as the focus of radical activism.[12] But it is greatly magnified in the online sphere, where an atmosphere of 'suffering, weakness and vulnerability' dominates. According to Nagle, this call-out culture became so oppressive in some online forums that the transgressive sensibility that once characterised the counterculture migrated to the political right, where a seething nest of racists and misogynists rebranded itself as the 'alt-right' and proceeded to troll progressive users in the most hateful terms imaginable. But it was less the moral vandalism of conservatives than the fact that *they too* began to regard themselves as *victims* of the progressive 'elites' that catapulted Trump and the other populists into power. So ineffective is the cybersphere at providing users with the recognition they need to build robust identities that even relatively privileged people come to see themselves as casualties, up to and including the great Man-Baby himself. Even Elon Musk, who has sought to make Twitter a global 'town square' where strenuous debate is permitted to all, whatever their ideological stripe, is more thin-skinned than a pinot grape in his interactions with fellow Tweeters, as likely to ban users for offending him as he is to readmit them for offending others. No doubt this is partly because he's a spoiled brat. But it's also, I think, because of the nature of social media, and of Twitter in particular.

There is another sense in which the internet and social media facilitated the populists' rise to power, which relates to the way technologies of absence affect the *shape* of human affiliation. As we saw in Chapter 1, one consequence of the invention of printing was the emergence of what Benedict Anderson called 'imagined communities'. Anderson was interested in how print capitalism led to the rise of modern nationalism and, by extension, the nation-state. To maximise circulation, print entrepreneurs would ensure their books were printed in the vernacular, as opposed to dead languages such as Latin and Ancient Greek. This led to a democratisation of information, as ideas and literacy spread throughout Europe, and systems that depended on traditional notions of divine right or hereditary power began to fade. It also meant that new identities began to form around 'national print languages', and these identities became increasingly important to the way individuals saw themselves. As Anderson put it, such identities 'are *imagined* because the members of even the smallest nation will never know most of their fellow-members, meet them, or even hear of them, yet in the minds of each lives the image of their communion'.[13]

It is interesting to think of this idea of imagined communities in connection to social media. One effect of this technology – a superfast field of symbolic production – is to increase the scope for intellectual contagion, and indeed *emotional* contagion, in a way that allows imagined communities to come into being with incredible speed. Consider the new form of activism where thousands of social media users will suddenly declare solidarity with some cause, as in the aftermath of the lethal attack on the editorial staff of *Charlie Hebdo*, which inspired the slogan '*Je suis Charlie*'. Or the way in which Islamic State or the pitiful subcultures of the alt-right make use of social media to spread their operations through the community. In the early 2020s, major cities across the globe were subject to semi-regular protests from libertarians and 'anti-vaxxers' opposed to COVID-related lockdowns and mandatory vaccination, and it is striking how 'cosmopolitan' many of them were. Some of the Australian protesters, for example, took to the streets in Trump-style red caps, or waved the Gadsden flag around like American revolutionaries. One man in Mount Gambier, South Australia, held up a sign declaring that mandates are an assault on 'our fourteenth-amendment rights'. Opinion is divided on whether or not he was able to find his way home unaccompanied.

Far from the horizontal organisation imagined by the tech utopians, then, social media would seem to be compatible with hierarchical and authoritarian movements – movements that in a number of cases bear more than a passing resemblance to fascism. The Italian sociologist Paolo Gerbaudo has argued that online political activism tends to default to charismatic 'hyperleaders' and a relatively passive 'superbase', and it's not hard to imagine how such a situation could turn even uglier than it has already.[14] The prospect of a future leader with a determinedly racist or misogynistic program, and a direct and deeply emotional relationship with his or her 'followers' in the digital space, is hardly a remote one. The thuggishness of a Rodrigo Duterte or a Jair Bolsonaro could just be the start.

The kind of political management that encourages a libidinal relationship between a leader and a leader's base is hardly a new thing in democratic politics: in the early twentieth century, Sigmund Freud's nephew Edward Bernays founded the field of public relations with a view to creating precisely that outcome.[15] But there is a sense in which social media seems particularly suited to the exploitation *and connection* of two forms of narcissism: the narcissism of the populist leader and the narcissism of his followers. Again, there are many factors involved in the populist catastrophe; but the role of technology is discounted at a cost.

Capitalism also clearly benefits from this political polarisation, and from the broader anxiety and dissatisfaction that underlies it. In order to ensure that the data they collect is as rich as possible, the platforms need to be volatile spaces – spaces in which no one can feel at ease. The entrepreneurs of Silicon Valley assert that their platforms are content-neutral – that they do not editorialise, or censor unpopular opinions. But what they have done to the public sphere is something far worse than to commandeer some bit of it for propaganda purposes. They have turned our conversation into a commodity and trashed the conversation in the process. They have monetised the tyranny of absence, and made political tyranny more likely as a consequence.

Even in the absence of tyranny, the effects of the internet and social media on politics should be a cause for concern. So large are the problems humanity faces that a resilient and productive public sphere is now an *existential* necessity if we aren't to simply cede control to technocrats or authoritarians, or to some grim combination of the two. Inscribed in the climate emergency, for example, is a mandate for concerted action and focus – for global solidarity. And yet, as the carbon economy trashes the conditions for life on Earth, the silicon economy trashes the capacity for intelligent and prolonged reflection that we need to rescue what remains of it.[16] The two forms of pollution are linked. Don't be fooled by the progressive noises made by Zuckerberg and his techno-chums. Just because Glencore and Google come from different parts of the forest doesn't mean they won't combine to burn it down.

You Are Here

How should we navigate this territory, bearing in mind that it is, for now, the *only* territory navigable? If social media is here to stay, at least for the foreseeable future, how should those of us who want to see a radically different kind of world, and who are interested in ideas, engage with it?

First, we need to put to rest the notion that there is any necessary relationship between the platforms and democracy. Happily, this process is already underway. As the 2011 protests faltered and Libya and Syria descended into war, mainstream commentary around social media took on a cooler, more balanced tone.[17] Social media, it was said, was not the driver of the unrest, but a tool that could be used to coordinate it. Moreover, it was becoming clear that the social network could also be used by authoritarians to shore up power in the face of such upheavals, through tried-and-tested tools of statecraft such as censorship and propaganda.[18] The Cambridge Analytica scandal, which began to break in 2015, set tech utopianism back even further. Far from being an algorithmic Excalibur, social media was now regarded as a double-edged sword.

Of course, there are a number of movements that couldn't have spread in the way they did had it not been for social media: the climate 'strikes' inspired by Greta Thunberg and the Black Lives Matter movement, for two. Like the uprisings in Tunisia and Egypt, the latter movement coalesced around instances of *visible* police brutality, notably the 2020 murder of George Floyd in Minneapolis. Social media was used to spread the vision of Floyd's fatal arrest. But it is the determination of BLM protesters to materially confront oppressive institutions that defines it as a political intervention.[19] By coming together physically, protesters remind institutional power that the body politic *is composed of actual bodies*, and that the 'legitimate violence' on which the state's power is assumed to rest is often illegitimate in the eyes of those on the receiving end. Liberal handwringing about the destruction of property or police stations always misses this point: that however legitimate or illegitimate we consider particular targets to be, a protest, in the form of a rally or a march, is an implicit warning to institutional power to remember where true 'sovereignty' lies.

Technology and human beings in sync: a BLM protest in 2015

Second, and whatever their particular merit, we need to remember the principle contained in Socrates' thoughts on writing: that the relationship between the form and the content of a communication technology is never straightforward. The tech utopians came late to the point about social media and democracy, permitting Big Data to paint itself as the champion of the Arab Spring, due to their faulty picture of technology. Theirs was an instrumentalist view, which holds that technologies are neither good nor bad, except in the hands of good or bad people. Technologies are tools that serve our purposes, indifferent to the uses to which they may be put. A plough is a plough; an engine is an engine; an internet protocol is an internet protocol. When the bright young things of Silicon Valley opine that 'society' should decide how their latest innovation is used, they are channelling this instrumentalist view of technology: *Let the philosophers and politicians determine when and how to use these inventions. I'm just here to build cool stuff.*

We should be wary of this argument, since (as we've seen) it denies the extent to which technologies play a constitutive role in human affairs.[20] Social media may be very good at getting people into public squares – as it did in Cairo and downtown New York – but it also reconstitutes the public sphere in ways that may undermine the ambitions of those protesters over the longer term. Those wanting to use social media to advance non-mainstream political ideas need to be aware that technology is political, in the sense that it affects what Winner calls 'the form and quality of human associations'.[21] The historian of technology Melvin Kranzberg set out this principle beautifully: technology, he said, 'is neither good nor bad; *nor neutral*'.[22]

Ideally, this principle would come to dictate the way we engage on digital platforms. If one effect of social media is to confound the personal and the political, the dialogic and the non-dialogic, and to generalise a feeling of misrecognition – of being unfairly treated or ignored – then avoiding the personal register and focusing on *ideas* is crucial. Without wanting to be sententious about it, it is clear that many writers and activists who engage on social media are often not *engaging* at all, as Richard Seymour implies in his matchless description of the phenomenon he calls 'critique-by-quote-tweet': 'Holding aloft a specimen of a really degenerate opinion, we mock it for having the quality of being an opinion, which is that it gets something wrong. Inviting others to join in, we treat disagreement, not as constitutive of any society, but as malevolence, idiocy or the cry of the loser. It is to be settled by group humiliation, sudden orchestrations of mob fury, the stiletto-stab of sadism.'[23]

It is less the nastiness of such behaviour than the fact it is unproductive that makes online politics so dispiriting. Yes, the platforms have their poets: people who can distil a complex argument into a tincture of devastating reason, taking the limitations of the form in the spirit of a sonneteer or a haikuist. But more often than not, social-media posts are less propositional than visceral – word-emojis designed to trigger a response, rather than to make people think. Deprived of meaningful recognition, we settle instead for 'reputation' in the form of likes, retweets and so on, adopting a transactional model of political and intellectual exchange.[24] We are not to blame for this calamity, but nor are we completely helpless in the face of it.

In 'Beyond Techno-Narcissism', Winner asks, 'Who are we on the Internet?' As primary forms of association are supplanted by technologies of absence, and the solidarity that emerges from embodied presence withers as a consequence, this question becomes all the more important. Not wanting to come across as Luddites and cowed by the moral panics of the past, we have failed to take it seriously enough, or even, really, to ask it at all. It's time to set that right, and to take a leaf out of Socrates' book.

If only old donkey-lips had written one.

3

ON THE DANGERS OF SOCIAL DISTANCING

Technologies of Absence and Human Empathy

These are uncertain times. We're all in this together. We're here for you. According to YouTube 'creator' Sean Haney (Microsoft Sam to his channel's subscribers), these sentiments are the key ingredients of pretty much every commercial advertising campaign to make reference to the COVID-19 pandemic. To prove the point he has made a video that cuts together about twenty ads but could stand in for any one of them. Set to gentle piano music, it begins with shots of deserted streets and ends with people clapping from their balconies, with lots of faraway looks and images of familial togetherness interspersed between. Regimes of social distancing become the occasion for an outbreak of solidarity, with Apple, AT&T and Grubhub urging us to join hands across the network.[1]

Okay, this isn't total bullshit. In the weeks and months after the novel coronavirus began its island-hopping tour of humanity, there *was* a lot of solidarity around, much of it spread through social media and video conferencing platforms such as Zoom. The solidarity took many forms: heartfelt thanks from students to their teachers; karaoke renditions of 'What a Wonderful World'; high-school choirs and professional orchestras relocated to the digital sphere. Grateful that we could not just hear but *see* each other in real time, we took to our screens with a new appreciation for the relationships we had taken for granted. No doubt these fuzzy feelings will be one of the nicer things we look back on when the 'coronacrisis' is a distant memory: that time when we seemed to rediscover the initial promise of the internet.

Neighbours in the San Sebastián region, locked down on the day of the Spanish Cup final

Yet the feeling was edged with sadness. For as grateful as we were for the vision of our loved ones, stuck in lockdown or recovering in quarantine, these technologies reminded us of what we missed – namely, the physical presence of others. Though many remained grateful to be working from home, away from the strained sociality of the workplace, most of us missed our nearest and dearest, who seemed suddenly dearer for being less near. From loved ones dying alone in ICUs, to unattended funerals, to babies born in quarantine, to the ineffable melancholy of life in 'iso' – so vividly distilled by the comedian Bo Burnham in his Netflix special *Bo Burnham: Inside* – the COVID pandemic reminded all of us how central bodily presence is to our social and psychological health.

Indeed – and as Burnham appeared to intuit in his songs about FaceTime and Instagram, and as Haney's 'spots' from Uber, Apple, Facebook and Amazon confirmed – regimes of social distancing are not encumbrances to the info-tech sector.

They are the info-tech sector's business model.

So far I've focused on how new media tools not only fail to provide the 'ground' for healthy social interaction but also work to undermine it. In this chapter, I want to broaden the focus to technologies of absence in general. By technologies of absence I mean those technologies that remove the human element from social activities and situations that would once have occurred in the presence of others: not only communication technologies but devices and systems that increase the distance or decrease the contact between individuals. In Part III, I'll discuss the effect that some of these technologies have on our *creative* being – on our sense of personal agency and freedom. Here, however, it's the effect on our *social* being that interests me.

It's worth just pausing to reflect on how unique our current social arrangements are, in historical (and evolutionary) terms. For the vast majority of our history, we lived in small communities – first in bands of hunter-gatherers, then in semipermanent pastoral settlements, then in agricultural villages. We used tools, because we're tool-using animals, but our tools did not remove the need to live in physical community with others. Today, by contrast, our social lives are embedded in technology to an extent that brings completely new kinds of social relationships into being. So used are we to this situation that it doesn't strike us as strange or novel. But we are living through a profound transformation. Historically, this is unchartered territory.

The main driver of this transformation is the computer, which has introduced relations of absence into almost every area of social life. In just three decades, it has gone from being a clunky box one sat at to type a résumé to something that is always with us in our smartphones. (Recent studies suggest that the average amount of time spent online is now between six and seven hours per day, with smartphones accounting for about half of that activity.[2]) At the same time, the algorithms used in those smartphones have spread to other artefacts, from warehouse automata and customer chatbots to fast-food menus, self-service checkouts and all the other devices and systems that together add up to 'the Internet of things'. Even when we are together physically, we are often not together in spirit, as computers carry relations of absence into the very heart of meatspace – not just into shops and onto public transport but into classrooms and the family home. The result is that we are rarely ever completely present or completely absent. If intimacy is heaven and solitude is hell, we exist in a sort of social purgatory. As the social scientist Sherry Turkle puts it, modern humans are 'alone together'.[3]

In reminding us of the importance of presence, the pandemic lockdowns might have prompted a period of reflection on this modern purgatory. But if anything, Big Tech companies embraced the opportunities presented by coronavirus, couching what Naomi Klein calls a 'Screen New Deal' in the language of public safety, in an effort to 'extend their reach and power' to as many areas of life as possible. In May 2020, for example, former Google CEO Eric Schmidt was appointed to lead a government panel on New York City's post-pandemic development, with a particular emphasis on remote learning, telehealth and internet broadband. Meanwhile, Governor Andrew Cuomo enrolled Bill Gates and his eponymous foundation in the task of transforming the city's education system, again with new technologies as the focus.[4] As for the emerging industry in driverless or 'autonomous' vehicles, its people could scarcely contain themselves. As one CEO told *Mashable*: 'Humans are biohazards, machines are not.'[5]

Silicon Valley entrepreneurs have long dreamed of 'smart cities' in which cybernetic technologies are used to control energy distribution, traffic flows, lighting, public transit and access to social utilities, and in which sensors, voice-activation systems and screens are mapped into every nook and cranny of civic life. In this vision of society, logistical decisions are made on the basis of near-perfect information, harvested from humans and non-humans alike, according to principles framed elsewhere, in the heads of info-tech billionaires – an arrangement, as US author and researcher Evgeny Morozov notes, that would effectively spell the end of politics, 'as it assumes the impossibility of wider systemic transformations'.[6] China's 'social credit' system, in which widespread surveillance is used to establish a trustworthiness 'score' for individual citizens, is condemned by most Western commentators as a sinister and illiberal development; but some versions of the smart city are scarcely less disturbing. In a leaked 2016 document, known internally as 'the yellow book', the Google-affiliated company Sidewalk Labs envisaged a reputational system that could serve as 'a new currency for community cooperation', rewarding people for good behaviour, keeping businesses honest, and so on.[7] As jaws hit sidewalks around the globe, the company moved into damage control, painting their thought experiment as the result of a too-eager brainstorming session.

Well, maybe it was and maybe it wasn't; but even if we accept their explanation, we are still left with a society in thrall to digital technologies – a society in which it is not unusual to spend more time in front of a screen than in offline conversation, and in which the processes of social life are increasingly mediated by technology. From delivery drones to autonomous weaponry, we seem to be moving ever further away from the embodied interactions of meatspace. Slowly but inexorably, a society based on relations of absence is replacing one based on relations of presence.

Can human solidarity weather such a trajectory? Or will the processes of disembodiment, under pressure from the priorities of profit and growth, undermine the social conditions needed for humans to flourish? Derived from the Latin word for 'solid' (*solidus*), the term 'solidarity' *implies* a physical human community, and yet communities are increasingly mediated, or even virtual, entities. 'Stay apart, stay together,' Samsung tells us in a COVID-era ad campaign, very close in tone to those in Haney's YouTube compilation. But what does 'together' mean here? When apartness becomes the standard, even the rule, what kinds of togetherness come to the fore, and where might they be leading us?

Technology in the Loop

The 'trolley problem' is now better known than at any time since the 1960s, when it was popularised by the philosopher Philippa Foot.[8] This is due partly to NBC's *The Good Place*, where it serves as the basis for some knockabout humour; but it's also due to the ethical dilemmas involved in the development of autonomous vehicles and other algorithmic machines. As a way to distinguish between different kinds of morality, the trolley problem is suddenly relevant, especially when it comes to new technologies that do not rely on direct human control.

In fact, it isn't a single problem but a series of related problems. By presenting respondents with a set of dilemmas that differ in some situational detail, philosophers endeavour to study the basis of humans' moral actions and beliefs, often disagreeing strongly about the significance of this or that set of responses. Sometimes, the only way to settle these disputes is to think up another version of the dilemma. The result is that the original scenario has now grown into a sub-discipline of moral philosophy. It even has a name: 'trolleyology'.

The basic form of the trolley problem, which is derived from Foot's famous essay on abortion, is usually just referred to as 'Switch'. In this scenario, a runaway trolley (or tram, if you're in the United Kingdom or Australia) is barrelling towards a group of five people, who will surely be killed when it reaches them. You, an onlooker, are standing by a switch that will divert the trolley onto a different track with only one person standing on it. Flick the switch and the trolley will change tracks, killing the one person but missing the five.

Diagram showing the original 'Switch' version of the trolley problem

Asked if they would flick the switch, most respondents reply that they would. A simple calculation is made – one death as opposed to five – and the moral decision formed on that basis. The decision is a utilitarian one, in that it follows the fundamental axiom set out by the philosopher Jeremy Bentham: that the morality of an action depends on its *consequences*. The moral action is the one that ensures 'the greatest happiness of the greatest number'.

But there's a fly in this ethical ointment. For when the details of the scenario are altered, changing the action but not the outcome, we tend to get very different results. In the 'Fat Man' or 'Footbridge' version of the problem, we are asked to imagine that we are standing on a footbridge, watching the trolley approaching the five people. Standing next to us is an overweight man, large enough to stop the trolley if we push him onto the tracks below. Again, respondents are asked to say whether they would sacrifice the one to save the five.

In this case, most people respond that they would not.

Why this discrepancy in response between scenarios? It depends on who we ask, but two popular answers are the 'deontological' one – that humans should be treated as ends in themselves rather than as merely means to an end – and the one derived from 'virtue ethics' – that pushing an overweight man off a bridge would be ruinous to one's character. Both of these responses hinge on the distinction between acts that *result* in the death of a person and actively *intending* that death in order to bring about a particular outcome. In the first scenario, a person is killed. In the second scenario, we kill a person. This distinction seems important to us.

But trolleyology isn't done yet. There is another version of the trolley problem that complicates the picture further. In the 'Loop' scenario, we are back at the switch, but this time the trolley will be diverted onto a sidetrack that *bends back* onto the track with the five people. The overweight man is on the loop – tied to the tracks, according to some versions – and we must decide whether to sacrifice him in order to save the five people on the main track. As in the footbridge scenario, that course of action would involve a choice to use one person as a means to an end; but in this case a far greater number of respondents report that they would sacrifice the fat man.

What explains *this* discrepancy? One possible answer – an intuitive one – is that it hinges on physical contact with the fat man, and on some deep sense that we do ourselves violence when we lay hands on another person with the intention of doing them harm.[9] This sense seems to bypass abstract morality: even imagining such a physical encounter induces in us a visceral response – one that goes to the embodied nature of human sociality. Yes, we can be a violent species; but acts of person-on-person violence are still rare enough to be profoundly shocking to most of us when they occur. In action films, barely a scene goes by in which people don't trade buffalo-stunning punches; but to witness a fight in real life, or any kind of physical altercation, is to be flooded with unpleasant sensations.

In other words, what distinguishes 'Switch' from 'Footbridge', or 'Footbridge' from 'Loop', or 'Loop' from 'Switch', has as much to do with our social *nature* as it does with our capacity for reason. In the end, it is easier to pull a switch than to push a person off a bridge, even when both actions have identical outcomes.

Or, to put it another way: hurting people is easier with technology.

I say our reaction to 'Footbridge' is visceral, but in fact there is good evidence to suggest that people, and young people in particular, are becoming increasingly utilitarian in their attitudes to *all* of the above scenarios. A 2017 study, for example, found that the appetite for 'utilitarian sacrifice' – deliberately killing one person in order to save five – was much higher among millennials than baby boomers, while a 2020 study analysing the responses of 70,000 participants in over forty different countries found that the average 'kill' rate for 'Footbridge' had climbed to 51 per cent.[10] Interestingly, it was higher in Western countries and lower in non-Western ones – a finding the authors put down in part to levels of 'relational mobility': in countries where people have more freedom to select interpersonal relationships, the number of fat men snuffed is greater, while in more traditional or conservative societies, where people tend to stay within their communities, the number of fat men snuffed is smaller.[11]

Why is the utilitarian mindset spreading so rapidly in modern societies? The authors of the 2017 study speculate that it has to do with a general decline in empathy, and the evidence suggests they have a good case: not only does lower 'empathic concern' correlate to utilitarian attitudes, it also appears that empathy levels have declined significantly in recent decades.

A 2010 study of US students even put a figure on this: it suggested that empathy declined by 48 per cent between 1979 and 2009, with much of that decline occurring in the last decade.[12] But as striking as these findings are, they don't really answer the question of why utilitarian attitudes appear to be spreading so much as reposition it: knowing that utilitarian attitudes are linked to lower levels of empathy still leaves unclear why empathy is dropping.

The finding that utilitarian attitudes are spreading faster in Western societies, where the social and psychological effects of neoliberalism are most pronounced, is surely part of the answer here. The dominant idea in such societies is that individual 'freedom' should take precedence over more communitarian forms of living. Such an ethos not only rewards those people who display non-empathic attitudes, but also engenders such attitudes more broadly.[13] (The fact that the Human Resources sector now re-imagines empathy as a sort of skill or competency – as in the 'empathy training' courses that disgraced politicians are now obliged to undertake – is one sign that it is on the wane.) But could it be that new technologies are also driving this phenomenon, and that these new technologies *in combination* with this predominantly Western idea of freedom are especially destructive of empathy? Remember that people are less likely to adopt a utilitarian moral framework when they imagine themselves pushing the man off the bridge than when they imagine themselves simply flicking a switch. And remember too that *all* technologies 'colour' the human ecology in ways that are not immediately obvious – they both affect and are affected by the political, social and cultural environment. As technologies of absence proliferate, we must take seriously the possibility that they are changing our moral subjectivities. The simple fact that some technologies disfavour certain human senses may be driving a fundamental shift in how we live in community with others.

Technologically advanced societies are increasingly governed by abstract relationships. And while no one believes that this will result in an outbreak of trolley-related deaths, it may be taking us towards the kind of society that runs counter to the social conditions necessary for a meaningful human existence – a world in which humans, in becoming less intimate, also become strange to one another over the long term.

Stupid about Humanity

The idea that technology may be changing our morality will not strike everyone as problematic. Utilitarian morality is still morality, after all, and 'Loop' doesn't say that we're becoming more unkind, only that we're becoming more utilitarian. For utilitarians, that's obviously good news: if people are calculating good and bad outcomes in the manner that Bentham recommended, so much the better for humanity. If empathy fails to maximise human welfare, why should we stick with empathy?

Whole books have been written in response to that question, both for and against the utilitarian position. But I think one salient objection to the latter can be stated simply and straightforwardly: taken as a moral system, *utilitarianism is stupid about humanity.*

Forget, for a moment, the many terrible acts that have been justified with utilitarian logic, from torture, to mandatory offshore detention, to the dropping of the atomic bombs on Japan; and forget as well the more obvious criticisms that are made of utilitarianism, such as the argument that happiness is impossible to calculate in the way that Bentham's ideas demand. Forget all that and focus instead on the *approach*, which ushers that ethos of calculation into the chaotic heart of human sociality – a sprawling emporium of the rational and the irrational, in which our abstract reasoning ability must mingle with those scruffier customers, *the emotions.* The Stoics saw reason as the highest human value; but if I'm disgusted by the thought of killing someone, then I think that tells me something important about the kind of thing I am and am not. Who knows, it might tell me something about the relationship between intimacy and a certain kind of order, or about the nature of altruism, or even about our altruistic nature. But utilitarianism is deaf to such speculations. In its prescriptive forms, it is a machine ideology – a monstrous ideology, even.

The philosophers who identify as utilitarians are morally serious people, eager to make the world a fairer place. I've no doubt they're decent human beings – generous tippers and kind to old ladies. They're not monsters. But the utilitarian attempt to separate human flourishing from the messiness of *being human* is bound to lead to monstrous outcomes. That we don't tend to run a series of equations before making moral choices is not a flaw or a failing: it is an important fact about our humanity (not to mention a precondition of our sanity). Utilitarian explanation, moreover, does appear to lend itself to what many would regard as 'inhuman' actions – actions taken by politicians, military planners and so on. Indeed, I would argue that utilitarianism is often a projection of what the philosopher of technology Lewis Mumford called 'authoritarian technics'.[14] By this Mumford meant those technologies and techniques that are conceived on an inhuman scale – indifferent to the social and creative needs of individual human beings. The ruthless efficiency of the assembly line is an example of this, in that it erases the humanity of those involved, regarding them merely as parts of the system. For Mumford, it was essential to scrutinise new technologies from the perspective of the 'organic human', whose basic needs did not change over time, or changed so slowly that they were objectively static.[15] The best technologies were developed, he argued, with those organic needs in mind, while the worst emerged out of other agendas: power, profit and the soulless ethos of maximum efficiency.

Mumford's concept of the 'megamachine' – a social system dominated by technology and functioning without regard for human needs – took him back to Ancient Egypt and the construction of the pyramids, an endeavour so large and ambitious that it gave rise to a unified worldview in which science, economics, technology and politics were regarded in machine-like terms. In similar fashion, the utilitarian worldview both reflects and reinforces the calculative ethos at the heart of technoscientific capitalism. Utilitarianism is so called because the things that produce happiness are said to possess utility; and utility, for Bentham, could be measured in money: if two different goods command the same price, it can be assumed that they generate the same utility for the purchaser.[16] Embarrassed by this early vulgarity, most modern utilitarians take a more sophisticated view of what constitutes utility, but they are still stuck with a moral system that renders human emotions and responses as *quantifiable*. The individual is reconceived as a calculating entity – a sub-system within a more complex system. Both self and society are reduced to algorithms.

Antonio Tempesta's 1608 depiction of the construction of the Great Pyramids

It is here that what Langdon Winner calls 'technological somnambulism' becomes explicable, as we are carried forward, zombie-like, towards ever more immersion in technology. Where pushback occurs, it is almost always on the basis that this or that technology may compromise privacy: confined to liberal habits of mind, we take our data to be something we own and its harvesting to be a species of theft. But it is not what we own but *what we owe to each other* that we really need to think about. We don't need a new set of terms and conditions. We need a new way of thinking about technology – as something that reconstitutes social life. In particular, we need to consider the possibility that technoscientific 'progress' is undermining the social conditions in which empathy and deep solidarity can flourish. We are moving towards the edges of the map, where the monsters breach on the yellowing parchment. We need some new coordinates.

Sleepwalking in the Smart City

Talking of somnambulism, here is the British novelist Ray Bradbury describing an experience he had in the 1950s, shortly after publishing his dystopian novel, *Fahrenheit 451* (1953):

In writing the short novel *Fahrenheit 451*, I thought I was describing a world that might evolve in four or five decades. But only a few weeks ago, in Beverly Hills one night, a husband and wife passed me, walking their dog. I stood staring after them, absolutely stunned. The woman held in one hand a small cigarette-pack-sized radio, its antenna quivering. From this sprang tiny copper wires which ended in a dainty cone plugged into her right ear. There she was, oblivious to man and dog, listening to far winds and whispers and soap-opera cries, sleep-walking, helped up and down curbs by a husband who might just as well not have been there. This was not fiction.[17]

Bradbury's horror will strike some readers as quaint, even as a little reactionary – the familiar harrumph of the technophobe who's mistaken novelty for civilisational decay. But with the Sony Walkman still twenty years away, and the iPod more than forty years away, Bradbury was ideally placed to recognise the social rupture such a device would represent if generalised to the population at large. The 'Seashell ear-thimbles' worn by the character Mildred in his novel at once annihilate distance and distance Mildred from those around her. This 'sleep-walking' state is now so ubiquitous that it does not strike us as unusual. But it is a revolution in our sensuous life, and in our sense of ourselves as connected to others.

The gizmo Bradbury spotted in Beverly Hills is the progenitor of the smartphone and its subsidiary technologies, which have revolutionised life for more than 80 per cent of the world's population. But the spread of absence doesn't stop there: it now reaches into every corner of society – into schools, into work, into private life. And while relations of absence are nothing new, algorithmic technologies represent a new level of penetration, and COVID-19 has accelerated the process still further. The Jerusalem (or Smart Jerusalem) dreamed of by the info-tech wizards is now on the horizon.

Take health and medicine, where algorithmic technologies are set to transform the carer–patient relationship.[18] Remote patient monitoring, wearable sensors, diagnostic algorithms, therapeutical apps and even telehealth are often justified with appeals to the bioethical principle of *primum non nocere*: 'first, do no harm'.[19] But what are the *hidden* harms of technologies of absence in this complex space? The Hippocratic Oath taken by medical students is named for the Greek physician Hippocrates, who also gives his name to the collection of medical works known as the Hippocratic Corpus. The Corpus is meticulously clear about how physicians should conduct themselves in the presence of the sick, touching not merely on professional etiquette but on what today we would call 'bedside manner'. Some of this content, such as the advice to physicians not to curl their beards like degenerate youths, has not dated well. But the Corpus is surely right to stress the importance of the *physical* relationship between carer and patient. Many studies have shown that simply going to the doctor can alleviate some patients' symptoms, and that emotional demeanour is an important factor in a patient's recovery from illness.[20] (According to one report, over 50 per cent of clinicians worry that telehealth hinders their ability to express empathy towards their patients.[21]) Where the illness in question is psychological in nature, the role of presence may be even more important, as face-to-face contact is often one of the things the patient is missing. When medicine is automated or moved online that sense of human connection is diminished.

As concerning is the effect that new technologies are having on the socialisation of students. Even before they begin school, most children are habituated to screen technologies, and some have failed to fully assimilate the pragmatic aspects of dialogue: allowing other people to speak, asking questions and seeking clarification, modifying language according to context and understanding physical cues that denote frames of mind and states of being, such as appropriate eye contact and body language.[22] Education itself is now in thrall to the computer, defended on the grounds that because computers are increasingly used in 'the economy', children should be using them too – an argument that is not only circular but also deeply utilitarian, subordinating the child's experience to a broader economic imperative. We are now beginning to move from computers in education to education in computers, as online schools and university courses proliferate in the wake of COVID-19. Researchers differ on whether such learning produces better educational 'outcomes' (another utilitarian word atrocity), but what every educator (including this one) knows is that the *experience* of education is greatly diminished. Denuded of the conviviality of the classroom, the experience becomes mere data transfer, teacher and student provider and recipient of information respectively. At the same time, the student experience is privatised – cut off from the opportunities for friendship, reciprocity and romance that characterise learning in traditional environments. The poet and teacher Robert DiNapoli captures the experience beautifully:

> Finding ourselves boxed into a participant's window in a Zoom or Microsoft Teams meeting radically unmoors us from all the conventional reciprocities of give and take between embodied self and embodied self we've been accustomed to for nearly all our lives. We can stare directly into the face of a relative stranger, who cannot sense our directness of gaze, which in former times would have constituted an uncomfortable faux pas. All the hitherto 'normal' protocols that governed the social exchange of looks have been put on hold. In effect we interact with simulacra of one another, which afford only thin ghosts of the personal presence such encounters in a shared physical space would entail. So are we 'there' or aren't we? Can we be truly present to one another in such technically mediated encounters?[23]

The answer, of course, is that we cannot, and that this experience of *disembodied presence* is the flipside of the *embodied absence* represented by Bradbury's sleepwalking woman and her contemporary equivalents.

In both cases, we are alone together.

One could go on endlessly in this vein, surveying the ways in which algorithmic machines reorder human sociality. But the key point is that such technologies now cut across many aspects of our social being, removing from the realm of human interaction a great deal of its emotional reward, and undermining empathic concern in the process. As the presiding geniuses of the animal kingdom, we are supple enough to adjust ourselves to such radically new relationships, but we do not do so without losing something – something important – of ourselves.

The tech sector itself is aware of this. 'Affective computing', which attempts to build emotionality *back into* algorithmic machines, is its solution. Pioneered by Rosalind Picard in her 1997 book of the same name, affective computing begins from an insight often overlooked in the development of AI – that emotion plays an important role in human reason and interaction.[24] Using signal-processing technology that can analyse, modify or synthesise signals in the form of sounds and images (speech-recognition technology is an example), it allows devices to 'read' human emotions as expressed through facial expressions, postures, gestures, intonation and so on, as well as the force or rhythm of key strokes or temperature changes in the skin. In the very near future, your car may be able to sense when you're distressed or angry, alerting other vehicles to your state of mind, and your mirror give you constructive feedback on a speech or important presentation. In one sense, the development of these 'smart' technologies can be seen as a branch of psychobiology, reading the body in order to discern the emotions and desires within it. No doubt it will prove extremely lucrative for developers and marketing executives alike, who already employ forms of 'sentiment analysis' (or 'emotion AI') on digital platforms, scanning our online textual leavings for evidence of our likes and dislikes.[25] No doubt it will be useful, too, in military and security circles, where 'emotional systems' that can identify targets and predict attacks have been mooted for some time. In 2009, for example, researchers at NATO presented an 'emotional neural network' model that could be programmed to mimic a 'trained human operator's glance' and weigh 'anxiety and confidence levels' before deciding whether to attack a target.[26]

Nevertheless, some commentators are optimistic about the rise of affective computing, suggesting that humans will come to form meaningful relationships with algorithmic machines. The novelist Jeannette Winterson, for example, looks forward to the day that 'helper robots' can be employed to keep elderly people company, and perhaps even record their memories for posterity. She's comfortable, too, with 'anthropomorphic sex robots', as long as we can agree to purge their design of 'backward-looking sexism and gender stereotyping'.[27] But while she is surely right to suggest that a '3-hole silicon-pornstar love doll' will reproduce misogynistic attitudes, her broader approach to human–robot relationships is incomplete, to say the least. For what we are asked to accept in this instance are machines that appear to understand our emotions, *but are not emotional beings themselves* – a 'social' relationship from which any genuine reciprocity has been removed.

'Ameca' – a lifelike robot created by Engineered Arts

The 'social' robot could in this sense deepen the very problem it was designed to solve. As Sherry Turkle argues, our growing dependence on robots of all kinds is linked to our decreasing expectations of people, and thus to the decline of empathy. 'The first thing missing in a companion robot,' she writes, 'is alterity, the ability to see the world through the eyes of another. Without alterity, there can be no empathy.'[28] The very idea of a caring robot is a contradiction in terms – one, moreover, that threatens to supplant a principal way that human beings become fully and functionally human: ministering to the needs of others. Winterson writes: 'There is no reason to believe that humans can only develop meaningful relationships with other humans.'[29] She's correct. From the 'transitional objects' that compensate the infant for its necessary separation from its mother, to the dogs and cats we keep as pets, it is clear that human empathy is not restricted to other humans, and is in some respects quite indiscriminate. But it's also clear that human relationships are qualitatively different to other relationships, and technologies that usurp those connections are potentially corrosive of them. Affective computing, in other words, would furnish us with yet another way to bypass all the messy compromises and challenges of embodied sociality, deepening the rifts that already exist in our atomised societies.

Something of this dynamic is at work in the phenomenon of virtual (that is, automated) companions, which use text and video interfaces to interact with users. The San Francisco–based company Replika, for example, makes AI chatbots that users can download as apps onto their phones or laptops, where they can customise their appearance and characteristics. Unlike Amazon's Alexa or Google's Nest, these chatbots are designed to interact personally with their users, drawing on previous conversations to ensure a certain continuity of relationship, conjuring the illusion of intimacy. The service is explicitly marketed as offering emotional support, with chatbots greeting their user-partners with comments such as 'How are you feeling?' and 'I've been missing you.' Though its basic service is free, Replika offers more romantic and sexualised functionalities to users who subscribe to its premium service – and many users experience feelings of intense attachment to their customisable partners. In February 2023, when Replika ran into technical problems as a result of an overnight software update, some users expressed feelings of genuine grief that their chatbots had lost the 'personalities' that had been developing up to that point. The ELIZA effect – a phenomenon coined in 1960 by the computer scientist Joseph Weizenbaum to describe the tendency of computer-users to project human characteristics onto their machines – had entered a fascinating and dangerous stage, in which users were encouraged to bond with algorithms constructed entirely from their own data and desires: a fundamentally narcissistic model that recasts companionship as calculation, quantification and informatics. On the whole, the mainstream media overlooked this crucial point, focusing instead on the character of individual avatars, Replika's treatment of its customers and the prospect of inappropriate content being made available to children. In what amounts to a different kind of projection, it missed the fundamental effect that artificial intelligence is having on human emotionality.

Such is the zombie ideology of technological somnambulism. Rather than thinking about how the technologies we have can be integrated into the relations we value, we greet the new generation of technologies as better than the last *by definition*, with perhaps a few caveats as to user privacy or the effect of adult content on kids. The info-tech guru takes to the stage, radiating positivity, and unveils his latest innovation – a smart speaker, perhaps, that can tell when its user is experiencing psychological distress. Or maybe he wants to share with us his vision of the networked city – of ubiquitous computing so sensitive to our 'needs' that the greatest happiness for the greatest number becomes a material possibility. As the representatives from IBM, Cisco Systems, Siemens AG and countless start-ups applaud, we conclude that he must know what he's doing, struck though we are by the conspicuous absence of human beings in his dazzling slides.

Game of Drones

Should anyone think I am overstating the revolutionary effects of algorithmic machines, let's consider developments in weaponry, which are beginning to transform the character of military conflict. Even as Russia bombards Ukraine with 'conventional' weapons, conflicts (one can scarcely call them wars) in which only one side is actually present on the battlefield are now an emerging reality. Earlier I said that hurting each other is easier with technology. In the military space, that proposition is being tested in real time.

In an article on lethal autonomous weapons (LAWs), AI researcher Toby Walsh calls for an urgent moratorium on the development of military hardware that operates without human oversight, referring the reader to an open letter with over 30,000 signatures, including those of some 5000 robotics researchers.[30] Invoking the trolley problem and the light it sheds on the complexities of human decision-making, he suggests that what the media likes to characterise as 'killer robots' should be banned by the international community.[31] In his letter, Walsh spells out the dangers of such autonomous weaponry, which could be used to seek out and kill combatants who meet predefined criteria, as well as for terroristic purposes. Such weapons represent a breach of trust, he implies: just as chemists rallied to the cause of a ban on chemical weapons after World War I, and many physicists supported treaties banning space-based nuclear weapons, so AI/robotics researchers should denounce the use of LAWs.

Walsh's campaign is important and just: the idea that a cybernetic machine could be programmed to take a human life and set to work autonomously is a profound departure from what most would regard as the basic ethical common sense that should obtain in the AI/robotics space. As long ago as 1942, the sci-fi author Isaac Asimov set out his Three Laws of Robotics, at the heart of which was the idea that a robot should never harm a human or, through inaction, allow a human to come to harm, even at the risk of breaching one of the other laws. Writing as Alan Turing and his team were cracking the Enigma codes (and building the first modern computer in the process), Asimov was not naive about the uses to which 'intelligent' machines might be put. But the idea that a robot could be programmed to kill a human would have struck him as monstrous.

But the 'third revolution' in military technology (after gunpowder and nuclear weaponry) is not confined to autonomous drones, and I wonder if, in one respect, Walsh is being too optimistic about our ability to control the technologies in our charge. Drawing a distinction between LAWs and 'remotely piloted drones for which humans make all targeting decisions', he says he wants to keep humans 'in the loop'. Yet as we know from our own incursions into trolleyology, much depends on the kind of loop we are in. The instrumental view of technology says that it is *us* who are in control, and that it is only our motives that matter. The reality is that technologies change us – they shape our subjectivities and colour our morality.

Nowhere is this more apparent than with military technologies, where innovation is a strategic necessity, and increasing distance between adversaries one consequence of that innovation. From the arrow to the gun to the intercontinental missile, weaponry has increased in range and sophistication in ways that have transformed the nature of conflict, as well as the subjectivities of those charged with fighting the enemy. To put it bluntly, increasing distance between combatants makes it easier for them to kill one another. This point is so obvious it is often missed – though not by the military planners, who invest time and treasure in inoculating their soldiers against the 'moral injury' that results from taking lives or simply witnessing acts of violence. Although David Grossman's famous study of non-firing soldiers in World War II is not without its (credible) critics, the problems it identified – the reluctance to take another's life and the psychological 'fallout' from acts of violence – are taken seriously by modern militaries, which have employed desensitisation techniques (human-shaped targets, intensive drilling) and the ethos of group responsibility in an effort to overcome them. Bertolt Brecht's line that every tank has a 'defect' – its driver – here comes into its own: humans are a problem militaries need to solve.[32]

As we move deeper into the digital era, these issues of physical and psychological distance are reaching an inflection point. For years before the US withdrawal from Afghanistan in 2021, the go-to military technology used by US forces was the unmanned drone. 'Piloted' remotely from bases in the United States (using skills perfected in computer simulations), such hardware leads to a form of warfare that is radically asymmetrical – in which the risk is to one side in the conflict only. The challenge of human presence still exists: the unmanned drone is a more intimate weapon than a long-range missile, in that it brings its pilot into visual contact with the people being targeted.[33] More broadly, however, the unmanned drone represents a major shift in soldiery, as the pilot becomes a mere node in the system – subject to moral injury, yes, but not to physical retaliation, or indeed to any physical danger at all. War becomes an abstract exercise, as a systems-thinking approach to conflict takes over, and residual notions of 'the warrior's honour' are jettisoned as a consequence.

Such systems thinking in the military space did not begin with unmanned drones. After World War II, in search of ways to deal with the prospect of war with a nuclear-armed Soviet Union, planners conceived the very technologies now incorporated in algorithmic machines. Faced with the prospect of nuclear annihilation, they imagined a decentralised system that would respond automatically to foreign attacks, while keeping communication running in the event of an outbreak of hostilities. In the early defence and warning systems of the 1950s and 1960s, scientists and engineers integrated information technologies into military command-and-control systems, developing the nascent disciplines of cybernetics and AI, and laying the foundations of the internet. At the same time, military strategy took on a weirdly abstract quality. Game theory, in which human decisions are seen as strategic interactions, with all 'players' trying to maximise their advantage, was used to build predictive models of possible scenarios, all of them based on the understanding that the actors within them were rational and calculating. As every human being on the planet became, in effect, a frontline soldier, humans themselves seemed to disappear from the military planners' considerations – an irony caught in Stanley Kubrick's movie masterpiece *Dr Strangelove* (1964), in which an exasperated US President reprimands his colleagues for fighting in the war room. What the novelist Martin Amis called the 'megadeath intellectuals' were in charge, and abstract 'war games' all the rage.[34]

War-gaming at the US Naval War College in 1958, using the newly installed
Navy Electronic Warfare Simulator

The new generation of military technologies is often contrasted favourably with non-tactical nuclear weapons of the kind that would lead to the scenario known as mutually assured destruction. But what the new military thinking inherits from the old is this highly abstract, systemic approach. It is 'network-centric' – founded on a range of integrated technologies, from GPS to wireless computers to remote physical monitoring of soldiers in the field. Yet what is consistent is the difficulty of distinguishing war from its simulation.

Such a scenario is one in which the notion of war as a battle between adversaries, and as morally distinguishable from other forms of violence, will become increasingly irrelevant. In its place is emerging a process of *subordination* characterised by targeted strikes – strikes that are a lot less 'surgical' than the military brass is prepared to advertise. Even now the United States runs many hundreds of drones over Afghanistan, the Middle East and Africa, often employing pattern analysis to identify and take out targets deemed suspicious enough to be expendable, and 'double-tap' strikes, which aim to kill first responders.[35] On the one side there is abstraction and absence, on the other rubble and broken bodies. Like the levitating, tech-obsessed island of Laputa in Jonathan Swift's *Gulliver's Travels* (1726), which keeps order in its colonies with threats to rain down death from above, technologically advanced societies will simply use their technological superiority to keep their enemies on the run, and largely invisible from the broader public. In the early 1990s, the philosopher Jean Baudrillard was widely mocked for declaring that 'the [first] Gulf War did not take place'. But Baudrillard did not mean this literally. He meant that what the military and media *presented* as the war was a spectacle, and a travesty of events 'on the ground'. (Joking with journalists in his famous briefings, the US general Norman Schwarzkopf was fond of showing smart bombs striking their targets – a phenomenon that led one commentator to remark on the way the screen, upon impact, 'conveniently destroys itself'.) In the era of the algorithm, however, there is not even a spectacle to contemplate, as conflict becomes a game of drones, played in our name, in places we won't visit, and for stakes we had no hand in setting – truly, a tyranny of absence.

Another Stanley Kubrick moment, this time from *2001*. An early hominid discovers that a bone can be used as a weapon against a rival tribe. The bone represents technology – a point the director makes explicit when the hominid throws it into the air and the movie cuts to a satellite. Kubrick's original idea in this sequence was to connect the weaponisation of the bone and the potential weaponisation of space, though he decided against this. Even as it stands, though, *2001*'s famous match-cut scene says something about the relationship between human beings and technology – that the very thing that makes us extraordinary is also the thing that threatens to destroy us.

That's never more relevant than when talking about those tools we call weapons. As the philosopher Theodor Adorno put it: 'No history leads from savagery to humanitarianism, but there is one leads from the slingshot to the megaton bomb.'[36] But as rational and creative creatures, we also have a *choice* about how we use tools, or whether to use certain tools at all, and the task is now to recognise that choices need to be made, and quickly.

Rather than simply acquiescing in the penetration of algorithmic machines, it is necessary to ask some basic questions about what problems, if any, such machines are solving and what their unintended consequences might be. I'm not suggesting that computers have no role to play in urban planning or the caring professions or education; but the systems thinking that sets the imperatives of economic 'efficiency' above our creative and emotional needs must be resisted strenuously. What Mumford calls 'democratic technics' has little to do with the institutions of government; it implies the subordination of technologies to what he calls 'the human scale' – an outlook born of the understanding that human beings' use of technology is both a means to an end and an end in itself, not merely a way of meeting some goal but a way of living out our humanity. As algorithmic machines become ever more powerful, something like this understanding of humanity's proper relationship to its tools becomes a political necessity. Upon pain of becoming uncanny to one another, we need to put presence back on the map.

After all, we're all in this together.

CODA

The Atomic Moment

On the morning of 6 August 1945, less than a month after the Trinity Test in New Mexico, three B-29s appeared in the sky above the Japanese city of Hiroshima. The city's residents paid them little attention: the sight of US reconnaissance planes was not unusual at this stage of the war. Mindful of the firestorms that had ripped through Tokyo and other cities due to US aerial attacks, the authorities had set some eight thousand schoolchildren to work preparing firebreaks. No siren sounded, and they returned to work. Only a few of them saw a large parachute unfurl beneath one of the planes as it turned around and pointed its propellers away from the city.

The use of atomic weapons against Japan – an action taken in the full knowledge that the damage inflicted would be almost entirely 'collateral' – is the moment in human history in which the technosciences meet human destructiveness at its lethal – no, genocidal – worst. The crime was not technologically 'determined': strategic intimidation of the Soviet Union and anti-Japanese racism were more proximate drivers, as was the moral degeneration that necessarily occurs in wartime. But its utilitarian justification, which centred on an argument about the number of lives that would be lost in the event of a ground invasion, bespeaks a new, and newly callous, willingness to turn technology to inhuman ends, and a new order of utilitarian calculation. As energy from the massive blast – several million degrees centigrade at its hottest point – travelled outwards at the speed of light, turning human beings within a half-mile radius into instant carbon statues of themselves, the human species was itself transformed, psychologically and morally. Not before or since has a strategic military decision been justified with a more bizarre 'moral' arithmetic.

In that sense, the atomic moment goes to the issue addressed in this first section: the fraught relationship between technology and human relations under pressure from a utilitarian mindset. (In a grim foreshadowing of the trolley problem, the second bomb was codenamed 'Fat Man'.) As noted in the introduction, however, the development of atomic weapons goes also to the issue of how the technosciences afford a new relationship with nature – one in which human beings are able not merely to harness or tame natural forces but to reconstitute them at the most basic level, in a way that lifts technological endeavour from the pragmatic to the truly Promethean. In splitting the atom, the Manhattan Project scientists did not use their understanding of nature to refine or re-imagine existing technologies; they based an entirely novel technology on a new, and abstract, understanding of nature – one increasingly applied to human beings themselves.

It is this second picture, and its consequences for humanity, that is discussed in Part II. But it is important to note that these two processes – the fragmentation of human society and the fragmentation of nature – are related. The highly instrumentalised view of life that emerges from the technosciences leads at once to a picture of society as composed of competing individuals, and to a picture of the individual as a basically machinelike entity. Social atomisation and physical atomisation are mutually reinforcing phenomena.

With that in mind, let's turn to the question of how the technosciences propose to remake humans at the level of embodied being.

PART II

HUMAN RESOURCES
Technology and the Body

Now I am ready to tell how bodies are changed
Into different bodies

TED HUGHES (*after* OVID)

4

HACKING HUMANITY

The Body as Machine

Nothing dates as quickly as a vision of the future. From flying cars to glass-domed cities to nuclear-powered vacuum cleaners, history is littered with duff predictions and ludicrous extrapolations. My brother-in-law, an award-winning architect, likes to begin his presentations with antiquated images of the City of the Future, as a warning to planners and developers not to get too carried away with their own sense of possibility.

A 1931 image depicting the futuristic *Ocean Express*, predicted to be able to make the run from Hamburg to New York in forty hours

It's less the technology than the aesthetics that amuse him, I think: the fact that the airships zipping across the sky are piloted by chaps in tweedy suits, or outfitted with Klaxon horns. Watching an old science-fiction movie often has a similar effect, as the characters wrestle with whatever calamity has befallen twenty-first-century Earth in gear suited to a glam-rock disco circa 1975. No less than the jetpacks and the bases on Mars, those shoulder pads and silver onesies consign our images of the future to the past.

The reason for this incongruity is simple: our visions of the future are coloured by the present. The British philosopher and critic Terry Eagleton makes this point in relation to utopias, which he describes as the 'most self-undermining of literary forms ... epiphanies of the beyond which bear witness to the fact that we can never attain it'.[1] In rebelling against the unimaginativeness of the present, such futurism simply reproduces it. In Sarah Scott's *A Description of Millennium Hall* (1762), utopia is imagined as a country estate where dwarves play harpsichords and tend to the shrubberies, while the hero-explorer of Francis Godwin's *The Man in the Moone* (c. 1620) climbs into the stratosphere on an elaborate kite pulled by twenty-five swans.[2] In these texts, and in many others, it's as if the human mind keeps trying, and failing, to achieve escape velocity.

Today, however, we find a new kind of vision creeping into the futurist fold – one that sidesteps the prediction problem by claiming that the future will be so radically different to the world we know now that it defies depiction. This is the tendency known as transhumanism, which asserts that current technological trends in AI, genetics and nanotechnology will mean that in the very near future humanity will merge with intelligent machines and expand its capabilities beyond anything currently conceivable. The fleshy sacks we call our bodies will be accessorised with hardware and software, and our minds uploaded to powerful computers, while medical progress will become so rapid that it outpaces human disease and ageing and brings immortality into prospect. But as for what our lives will be *like* under this new transhumanist system: one may as well ask what it's like to lay eyes on the entire electromagnetic spectrum. It simply cannot *be* imagined.

Part science, part philosophy and part religion, transhumanism began in the 1950s as a social and scientific movement stressing the planned improvement of humanity. Its contemporary advocates take science and technology to be the principal forces in human affairs, often placing them within a cosmic and evolutionary frame. According to this view, the story of humanity is the story of how the energy flows that began with the Big Bang were redirected via technologies of ever-increasing efficiency, to the point where they allow the human species to take control of its own evolution. Transhumanists tend to regard humanity's natural and technological evolution as part of the same overriding process. What is the in-principle difference, they ask, between an ancient pot or a rudimentary plough and an implant that can connect your brain to the internet? Are they not aspects of the same drive towards mastery that defines *Homo sapiens* as a species?

A small number of transhumanists are brave enough to put their micro-processors where their mouths are. The artist and activist Neil Harbisson, for example, who was born completely colourblind, has an antenna attached to the back of his head that allows him to experience colour in the form of 'audible vibrations' through the skull.

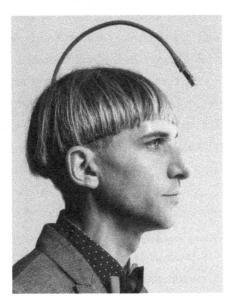

Biohacker and artist Neil Harbisson with his cyborg antenna implanted in his skull

Or take the British engineer Kevin Warwick. In 2002, he arranged for a device to be inserted under the skin of his left arm and 'interfaced' with his nervous system, which was then linked to an internet node at Columbia University. This allowed him to remotely control a robot arm in another country, and even to obtain feedback from its fingertip sensors. Extending the experiment a few years later, and having made a few adjustments to his wife, Irena, Warwick then demonstrated that an electric signal could be sent directly from one nervous system to another. Whether he'll be able to take the next step and engineer *telepathic* contact between himself and Mrs Warwick, opinion is divided. But that is the engineer's stated ambition. And they say romance is dead.

These madcap projects to one side, what transhumanism presents is not a program for the future but a teleological vision of humanity as moving in a particular direction, in accordance with its greatest gifts. Humans will merge with machines, it says, because they are machine-makers. Some of its advocates set this teleology within a conventionally humanist frame, as when the biologist Julian Huxley, considered by many to be the founder of transhumanism, argued in the early 1950s that social policy and eugenic science should be geared to improving humanity in line with enlightened ideas of progress – not by supplanting Mother Nature, but by picking up where the old girl left off. Others take the 'post-humanist' line that new and emerging technologies will turn us into something else entirely. For the US futurist Ray Kurzweil, for example, the exponential rate of technological progress, or 'law of accelerating returns', means that humanity will cross a threshold into a radically different mode of existence by 2045. Thus, human beings, who are already pretty complex, are poised to become more complex still, propelling us to what Kurzweil calls 'the Singularity' – 'a future period during which the pace of technological change will be so rapid, its impact so deep, that human life will be irreversibly transformed'.[3] Not only will artificial intelligence pass a threshold into artificial *general* intelligence – into artificial consciousness, more or less – but human beings will become *super*-intelligent, uploading their minds to powerful machines that extend their capabilities a billion-fold. Never mind time travel, or underwater cities, or all that retro-futurist kitsch. The *telos* – the purpose – of the human species is to liberate itself from the messy business of being (physically) human at all. All within the span of a millennial's lifetime, if Kurzweil is to be believed.

It is tempting to dismiss such ideas as fantasy. Tempting, and often necessary: there are many fundamental objections to the ideas formulated under the rubric of transhumanism, including the objection that since modern computers in no way resemble human brains it is premature to go around saying that the two are heading towards a material synthesis. But transhumanism's broader approach to the future – its sense of where our species is heading in terms of its relationship to technology – is not so easy to discount. Not because it is right to imagine that we are bound by fate to escape our bodies, but because its central idea that technology is the measure of humanity's progress is now deeply ingrained in 'advanced' societies – i.e. societies no longer organised around traditional systems of custom and belief. Even if the transhumanists are wrong to imagine that computers can think like human beings, or that human beings can be reprogrammed like computers – and I think they almost certainly *are* wrong – their relaxed approach to transformative technologies is still a symptom of a broader faith in new and emerging interventions – a faith that could have, and may already be having, profound and profoundly troubling consequences.

It is, then, less Kurzweil's Singularity than the 'soft' transhumanism of which it is a symptom that should concern the student of technological trends. Yes, the transhumanists can appear quite batty, but both the military research establishment and Big Tech appear perfectly happy to fund their battiness. Kurzweil himself has been employed as Google's director of engineering since 2012, hired by Larry Page to 'bring natural language understanding to Google' (natural language processing allows computers to recognise text and speech). The appointment was logical. Google is determined to integrate humans and machines as far as possible, through its expanding range of 'smart' products: glasses, headsets, speakers and so on. One doesn't need to buy into the idea that we are destined to abandon our bodies to recognise that embodied life is increasingly colonised by smart machines, and that this development may have consequences for our social condition that we have barely begun to wrestle with.

In the technosciences more broadly, we find the transhumanist impulse alive and well in the great convergence of info, bio and nanotech, which together allow us to intervene in the smallest elements of nature. The development of CRISPR technology, which we'll consider at length in Chapter 5, is hugely significant in this regard, as we can now edit strips of DNA in a way that allows us to modify organisms. Culturally, too, we are getting used to the idea that new medical technologies can play a liberating role in human affairs. For example, IVF clinics are beginning to experiment with 'three-person IVF', where mitochondrial DNA from a healthy egg is inserted into an unhealthy one in an effort to prevent mitochondrial disease from developing in newborns. Many would point out that human beings have always used technology to harness nature. But there is a big difference between harnessing natural forces and reconstituting nature itself. That is a far more radical proposition – the result of a major shift in perspective – and one that an overriding faith in technology and progress has caused us to overlook.

In one sense, transhumanism is as much a projection of present obsessions as Godwin's swan-powered microlight or Bertie Wooster's flying jalopy – less a viable proposition than a reflection of our attitudes towards humanity, technology and the relationship between the two. As the philosopher Bob Doede suggests, its 'buoyant eschatology' is founded on a wrongheaded idea of the kinds of creatures humans are, and it is this, not Kurzweil's dream of a digitised humanity, that is pregnant with catastrophic potential.[4] In love with the idea of technological progress, we accept the transhumanists' central argument: that an ever-deepening symbiosis of smart machines and human beings is simply part of a trajectory that began with mastery over fire. Like Viktor Frankenstein, we have crossed the line separating scientific curiosity from Promethean hubris, as scientists propose to 'pioneer a new way, explore unknown powers, and unfold to the world the deepest mysteries of creation' while we either cheer them on from the sidelines or accept the resulting disruption as the inevitable price of progress.[5]

Part I of *Here Be Monsters* considered the effect of what I called 'technologies of absence' on our social and emotional lives. Part II focuses on issues of *embodiment:* on how the modern technosciences promise, or threaten, to remake humanity at the level of its physical being. Coming to the fore at a point in history where a particular, highly questionable, idea of human freedom and flourishing dominates, how might such bodily interventions affect human beings in the short- and long-term? Will we continue to recognise ourselves in others, or will social solidarity yield to a widespread sense of alienation?

In order to address these questions – and to make the case for a moratorium on some of our most Promethean endeavours – I'll consider a range of technologies, including genetic engineering, body modification and mood-altering pharmaceuticals. Before we get to the technology, though, let's examine the view of humanity on which the technosciences are founded. It is necessary to name the spectre – the monster – that already haunts our imagination in order to begin the exorcism.

The Clockwork Universe

The version of humanity that began to emerge in the early modern period now dominates our conception of self in ways that may not be immediately apparent. In the past, most religions and systems of morality distinguished between the body and the soul, granting human beings an existence beyond the physical world. But in recent times, we have begun to think of humans in material terms, bound by the physical laws of nature. This naturalistic view of humanity – while it is by no means universally accepted – has liberated human beings from many of the tyrannies associated with religion, and has also led to vast improvements in health, production and understanding. But it brings with it tyrannies of its own, including a vulgar reductionism in which we are no different from the machines we have created. Combined with neoliberal capitalism, this vision of 'the human machine' now threatens to undermine human nature itself.

The emergence of this notional monster is bound up with modern technoscience, the roots of which lie deep in the soil of that great efflorescence of astronomy, physics, mathematics and anatomy known as the Scientific Revolution. From around the middle of the sixteenth century, the religious and philosophical traditions that had dominated the Middle Ages began, very slowly, to yield to a new outlook based on the 'scientific method', in which knowledge derived from the senses (empiricism) was combined with 'inductive reasoning' (using individual observations to establish general principles). Emphasising abstract reasoning over common-sense interpretation, and a quantitative approach to the universe rather than a qualitative one, this outlook represented a move away from the ancient search for 'final causes' – for the *purpose* of phenomena – and towards material explanation. It was not the *why* but rather the *how* of natural phenomena that concerned the new science. Breaking its ancient links with the sacred, science became a separate discipline with a distinctive view of the universe.

As the insights of empirical science shaded into the Enlightenment towards the end of the seventeenth century, both scientists and their intellectual champions in the new 'mechanical philosophy' began to regard the natural world as itself a gigantic apparatus – as a vast domain of matter in motion governed by material laws. The physical universe, this approach suggested, was much like a clock whose separate mechanisms were set in train by direct physical impact. It followed that the job of the scientist was to understand how the different mechanisms of the universe interacted – an approach that led (again, slowly) to a widespread disenchantment of nature.[6] Despite many scientists continuing to believe in God and magic, Enlightenment rationalism had the effect of emptying the world of mystery. Such ghosts as remained in the universe-machine were living (or haunting) on borrowed time.

A mechanical model of the solar system, similar to the one depicted in Joseph Wright's
A Philosopher Lecturing on the Orrery (1766), below

Looking back on this revolutionary period, the German philosopher
Martin Heidegger (1889–1976) presented it as both a cause and a symptom
of a particular approach to technology. Before modern science, Heidegger
suggested, technologies simply drew on nature's pre-existing potential,
in a way that respected its intrinsic value. An old windmill, for example,
channels the wind's energy, but the wind is not transformed or used up in

the process. But as the normative and hierarchical cosmos of the medieval world gave way to the modern one, technologies took on a different aspect – involving active intervention in the world. Unlike a windmill, which is 'left to the wind's blowing', a factory or a coalmine 'reveals' the land as a resource to be exploited and converted into energy. The result of this is that nature begins to lose its intrinsic value. It is now seen as a 'standing reserve' to be hoarded like supplies in a kitchen, inventoried and pressed into service according to our demands. For Heidegger, this outlook is self-reinforcing: once we come to see nature in these utilitarian terms, we condemn ourselves to always see it so. Everything appears to us as a source of energy, as something we must organise, or harness, even to the point that we see *ourselves* as a means to an end, as raw material. We too become parts in the great machine of the universe.[7]

Heidegger did much to illuminate a certain understanding of self, but that understanding doesn't lead us directly to transhumanism. For human beings are rather less predictable than the movement of the planets or the laws of geometry. It follows that the rejection of God, and of the body–soul paradigm that goes along with most monotheistic religions, does not automatically 'reveal' human beings as machines, distinguishable from handlooms or calculators only by dint of their complexity. There are a number of other hurdles that need to be cleared before that view can take hold.

In particular, there is the so-called mind–body problem: the attempt to differentiate and to reconcile our mental and physical properties. If the universe is a material entity, and everything in it is material too, how is that I (a material being) can form what seem to be *non-material* representations of it in what I call my consciousness? What status do the images of stars and planets that came into my mind a few seconds ago, as I was typing the word 'universe', have within the cosmic apparatus itself? Is one a sort of subclass of the other, or are they fundamentally different things? And if they *are* fundamentally different things, how do they interact with one another?

The French mathematician and philosopher René Descartes (1596–1650) was one of the first Enlightenment thinkers to wrestle systematically with this problem, in a way that aimed to reconcile a mechanistic vision of the universe with a more conventional vision of the soul. In *Meditations in First Philosophy* (1641), he described two irreducible realms: mind, the essence of which is thinking; and matter, the essence of which is 'extension' in three-dimensional (physical) space. Having defined these separate realms, he then attempted to show how these 'substances' interacted within the body. Rather strangely, given our current knowledge about the physiology and functions of the brain, Descartes nominated the pineal gland as the site of this mysterious interaction. But the question of *where* the interaction occurred is less significant than the question of *how*. As Descartes' contemporaries were quick to object, how could an immaterial substance affect a material one at all?

Despite his failure to adequately think through the relationship between the brain and consciousness, Descartes' 'substance dualism' was the origin of the now-conventional distinction between consciousness and the physical brain, and started a centuries-long debate between philosophers, theologians, psychologists and scientists. Quite possibly it will never be fully resolved, and I certainly don't propose to adjudicate between the great interventions of Kant, Nietzsche and the rest of the philosophical bigwigs. It is enough to say that this is a problem that has exercised the finest minds in philosophy; that it has led to some really terrible ideas; and that it is one of those really terrible ideas that underpins the vision of humanity at the heart of technoscientific capitalism. This is the idea that the brain is a computer.

The Lovelace Test

Metaphors for the mind and brain are often drawn from the technological advances and artefacts of the day. Socrates and Plato, for example, likened the mind to an aviary, while Sigmund Freud compared it to a steam engine from which psychic energy is intermittently released. But the modern comparison between the brain and a computer is much more than a metaphor. As we saw in Chapter 3, it was the nuclear standoff between the United States and the USSR that led to major strides in computing, as the imperatives of command and control demanded algorithmic machines that could operate independently. But in attempting to develop 'thinking' machines, scientists also landed on a vision of the mind as essentially computational. In many respects, the brain and the computer became conceptual models for one another.

This mental-computational paradigm is one of the key features of cognitive science, a rich field focused on the nature and functions of intelligence in the broadest sense. Originating in the mid-1950s, it grew out of cybernetics, which sought to understand the mind in order to build intelligent machines, or machines that could respond to stimuli independently, as a human might. Cognitive science's founding idea was that the human mind is explicable in terms of information-bearing structures located in the physical brain ('representations') and the computational processes that operate upon those structures.[8] As the field has developed, the analogies between biological and non-biological systems have come to reinforce one another, to the point where new knowledge about the brain is employed to develop new kinds of computer, and breakthroughs in computer development suggest new complex models of the brain. The result is that contemporary cognitive science offers many different computational models of mind, and that modern computers are increasingly characterised by 'parallel processing' – they are capable of performing many operations at once, much as the brain appears to do.[9]

The comparison between the brain and the computer certainly makes intuitive sense. Computer programs do appear to be able to solve complex problems using algorithms, so perhaps it is reasonable to see human beings – problem-solving creatures, after all – as essentially computational. Not only do standard programming languages use sequences of 'IF/ THEN' instructions, which suggest a model of how people make plans ('IF I get a beer from the fridge and drink it, THEN I will feel relaxed'), but sophisticated computers are also pretty good at learning from their 'experiences' in the way that human beings do (beer-induced hangovers to one side). To this extent, it doesn't seem like a stretch to say that in certain crucial ways computers and human brains are alike. 'IF computers think, THEN human brains compute' is perhaps a logical way of putting it.

Yet there's a flaw in this algorithm. For what is the content of 'think' in this instance? Rather like the science-fiction writers who imagine that machines will achieve self-awareness at a certain level of complexity, the artificial intelligence boffins often reduce the concept of thinking to a question of sorting information. But as Ada Lovelace intuited as long ago as the 1830s, there is more to human thought that simple information-processing. Lovelace, despite recognising the revolutionary potential of Charles Babbage's Analytic Engine – widely regarded as the world's first computer, and a project to which she was central – saw that even future computers could never be said to actually *think*, if by thinking we mean the mental activity engaged in by the human animal. Her own thinking on the matter was elegant and simple: the Analytic Engine, she wrote, 'has no pretensions whatever to originate anything. It can do whatever we know how to order it to perform.' It was this view that Alan Turing targeted when formulating his Turing Test, which states that a machine exhibits intelligence if it can interact with a human being in a way that convinces the latter of its sentience. But as the original name of the Turing Test, 'the imitation game', suggests, that computer would not be displaying intelligence in the sense Lovelace identified, but merely the impression of it. The same goes for Alexa and her equivalents: since they cannot 'originate anything', they do not pass the Lovelace Test.

It is clear from the confusion that often surrounds the distinction between intelligence and consciousness that much contemporary 'cognitivism' 'has lost its bearings in the conceptual fog of its own abstractions and metaphors' (Doede).[10] As the chess grandmaster Garry Kasparov and the Go prodigy Lee Sedol can attest, computers can play complex games of strategy to a level more advanced than any human being. They can also fly aeroplanes, recognise faces and translate one language into another. But they do not possess animal consciousness, and the idea that they will develop it when they pass a certain threshold of 'intelligence' is poorly evidenced, to say the least. As the neuroscientist Christof Koch has argued in *Scientific*

American, the view that artificial consciousness (as opposed to artificial intelligence) is achievable is a category error. Mimicking the functionality of the brain is not enough to create consciousness, any more than simulating a weather system is enough to create cloudy skies or sunshine.[11] Raymond Tallis – another neuroscientist, as well as a brilliant philosopher – links the computational paradigm to the broader mind–body (or mind–brain) problem. He suggests that trying to understand consciousness by studying the neural activity in the brain is like applying a stethoscope to an acorn and expecting to hear the rustling of the forest. Attempts to explain an experience such as love through functional magnetic resonance imaging (fMRI), or to replicate it in computer models, are for Tallis always doomed to fail, since they can never account for intentionality. This is the power a mind holds to grasp something in a way that goes beyond the mechanical – that goes, for example, to my *experience* of a flower as opposed to the causal chain of events (light hitting a flower and entering the eye) that makes an image of it in my visual cortex.[12]

These objections have been widely aired in the decades since cognitive science was in its heyday. What looks, on the face of it, like a neat solution to the mind–body problem framed by Descartes is in fact a projection of the kind that Heidegger (writing in the 1950s, incidentally, just as cognitive science was hitting its stride) regarded as characteristic of a society that had begun to think technologically. We could even say that it *reframes* the problem in a way that replaces mind and matter (or body and soul) with software and hardware, and as such gets us no further along the road to explaining human consciousness.

Nevertheless, such informational dualism remains hugely popular, hegemonic even, in secular or 'advanced' societies. We can see this in the way we talk about our own cogitations as 'processing information', and in the way that very different notions of 'information' are swapped in and out, combined and confused, in discussions of consciousness. We see it too in the psychology books that invite us to 'rewire' our brains; in the models of AI developers and behavioural economists; and in the chunky histories of the human species increasingly favoured in Silicon Valley – histories that tend to present human beings as complex biological machines whose special gift is to channel the energies of the universe through technologies.[13] And, of course, we see it in transhumanism, in the idea that human beings will merge – will *physically* merge – with those technologies.

Such software–hardware dualism is part of a broader informational worldview in which everything is reduced to its smallest parts: to the neuron, to DNA molecules, to the atom. It is no coincidence that the computational theory of mind emerged in the shadow of the atom bomb, the development and deployment of which raised the curtain on a radically new relationship between human beings and the natural world – one based not on control but on *reconstitution*.[14] Like the scientists of the Manhattan Project, the cognitive scientists of the 1950s and 1960s were operating in a new and essentially reductive paradigm – a paradigm in which the smallest 'units' of a phenomenon are taken to be its essence. The British philosopher Mary Midgley described this new paradigm as follows:

> First, at the physical level, the idea of the atom was dramatised by bombs and by the promise of atomic power, so that the world seemed to consist essentially of atoms. Second, in human social life, computers emerged, and it was promptly explained that everything was really information. And third, on the biological scene, genetic determinism appeared, declaring that (among living things at least) everything was really genes and we were only the vehicles of our genes, but that (rather surprisingly) we nevertheless had the power to control them.[15]

That last point is central to the understanding of the technoscientific worldview. For reductionism is inseparable from the desire to manipulate the natural world in far-reaching and fundamental ways. The artist and author James Bridle suggests that the rise of computational thinking is closely related to the notion of 'solutionism': the belief that any given problem can be solved by the application of technology.[16] When the whole world is composed of information and data, what is to stop us from 'reprogramming' it according to our species' needs?

What, indeed, is to stop the species from reprogramming *itself*? As American academic Sheila Jasanoff has suggested, one of the dangers of modern genetics – or of the view of the world that often attends it – is that life devolves into 'just another object of conscious design, valued mainly for our ability to manipulate it, commodify it, and profit unequally from those acts of appropriation'.[17] Many of the books on modern genetics have titles that suggest the 'code' of life has been cracked or broken as a consequence of the field, while others adopt a religious register, referring to genetics as the 'book of life' in a way that accords a privileged role to those who work in that field.[18] Such language is both the cause and the symptom of a mindset that has come to see life as something to be engineered. Everything in the material universe, up to and including human beings, is cast as a kind of shadow-phenomenon of a code to which we have the key.

'They say it got smart, a new order of intelligence,' rasps Kyle Reese in *The Terminator* (1984), referring to the Skynet computer system that turned upon its human masters in the catastrophe known as Judgment Day. But such sci-fi gets the danger backwards. It isn't that machines have become so powerful that they threaten to destroy humanity, but that humanity has come to see life, the universe and everything in machine-like terms. In thrall to a technologised vision of the world, we have merely replaced Cartesian dualism with a dualism of our own – one based on the great artefact of our time: the computer.

Atomisation versus Holism

One of the principal criticisms made of cognitive science, and of the computational theory of mind in particular, is that it overlooks the *contextual*, as well as the *social*, character of consciousness. In *Artifictional Intelligence* (2018), Harry Collins takes issue with Ray Kurzweil's claim that 'the kind of pattern analysis used by modern programs is the same as that used by human brains' on the basis that computers lack context-sensitivity.[19] Computers have neither the sense of themselves as existing in relation to a particular environment – as in my example of the flower earlier – nor the sense of themselves as one 'self' among many. They have neither intentionality nor shared intentionality, no notion that they exist in a community of other beings *with their own intentions*. Such shared intentionality not only distinguishes human beings from machines; it also distinguishes them from other animals. Take the simple act of pointing: a motion peculiar to human beings. If I point a dog in the direction of its ball, it reflexively looks at the tip of my finger. But a human being knows immediately that I am referring to something 'out there'. In this sense pointing is an intermediary stage between the world of our own sensuous experience and the capacity in which the social and creative elements of humanity combine most fully: the capacity not merely to communicate, as a bird might with a mating call or an ant with a tiny squirt of hormone, but to *name* the world, *to have a language*, in a way that allows us to represent not only things that *do* exist but also things that *might* exist. As Tallis puts it: 'We are explicit animals who *lead* our lives, living out shared and individual narratives, rather than merely *living* them, who are conscious of ourselves, of others and the material world, and of its intrinsic existence and properties in a way that no animal [and certainly no machine] is conscious.'[20]

In downplaying the social nature of consciousness, cognitivism makes for a tidy fit with neoliberal capitalism. In fact, there is a deep relationship between the informational view of life and the information economics that emerged in the decades after World War II. Capitalism has of course always been beholden to the concept of calculation, as goods and services are rendered as objects for trade through the pricing mechanism. But it was the information revolution that emerged from the early development of computers that caught the imagination of economists looking for ways to model the relationship between consumers and goods in a competitive market. This led to the emergence of game theory, in which individuals are 'players' who are assumed to make decisions in their own interests – a reductive view of human behaviour that nevertheless allowed economists to build predictive mathematical models in order to better understand markets and maximise efficiency. Increasingly, economic behaviour was seen in terms of rational interaction and information processing.[21]

The analogies between information systems and human behaviour in market contexts formed the basis of a new economic orthodoxy, based largely on the ideas of Friedrich Hayek, who by the 1960s had come to view thinking individuals as almost superfluous to the operation of the economy.[22] The best way to allocate resources, thought Hayek, was to leave decisions to the marketplace, which acted as a sort of omniscient information processor. This emphasis on information came to dominate economic thinking, as the 'rational actor' of game theory was eclipsed by a vision of human beings as essentially computational, less strategic players than responsive automata. The result was that economists began to focus less on the question of how markets could deliver what people wanted and began to focus increasingly on how businesses could maximise profits *regardless of what people wanted*.[23] While rational choice theory had assumed that humans make decisions on the basis of 'perfect information' and in their own self-interest, the new economics began from the assumption that humans were themselves informational, and thus subject to being hacked and reprogrammed. It was not rationality but the *limits* of rationality that excited these behaviourists.

The new economic individualism is thus one in which choice and individual endeavour continue to be lauded at the political level, while in the economic sphere, and certainly in the information economy, humans are seen in mechanistic terms – as aggregations of neurons and genes whose behaviour can be anticipated *and altered* through algorithmic prompts and nudges. In one sense, rational choice theory remains the ostensible story of how markets work; but capital's investment in behavioural psychology and neuroscientific models gives the lie to the ideology: it tells us we are demigods but treats us like laboratory animals.

Today these two visions are coming together in a way that brings transhumanism into prospect. Part of the neoliberal ethos is to transfer risk and responsibility from institutions to individuals: it tells us we are on our own and that it is up to us to make our lives better, to distinguish ourselves as individuals in the great marketplace of society. But what happens when this economic individualism mingles with a view of human beings as informational entities? In that case, surely, the move away from a communitarian emphasis and towards an individualistic one could engender a reversal in 'the direction of fit' between human beings and technology, as technologies that once held out the promise of building healthier, happier societies are employed instead to transform ourselves into *the kind of people we think we should be*.[24] If new technologies can make people 'happier' by making them prettier, or more distinctive, or by making their children more successful (taller, cleverer, more likely to secure high-paying professional jobs), then who are we to deny them that choice? Isn't it an individual's right to transform themselves in line with their own 'identity'? Thus two varieties of atomisation – the atomisation of society caused by neoliberalism and the atomisation of human beings suggested by the technosciences – are combining in ways that could well 'unground' human beings from their given nature.

Such technological interventions would almost certainly reproduce the therapeutic culture that has risen under neoliberal capitalism. Doede has suggested that this therapeutic culture has manured the ground for the transhumanist impulse by pathologising human dissatisfaction, while also proposing new interventions that 'cure' us of these putative ills. As we'll see in Chapter 7, such a progression is already discernible in the spread of antidepressants from those with conditions such as bipolar disorder to those who may be struggling to cope with the pressures of twenty-first-century life. Sadly the focus on depression and anxiety that followed the global COVID-19 lockdowns did not lead to any significant discussion about why those conditions appear to be rising *in general* across developed societies. That would mean thinking about those societies – about the design of communities, the nature of work and the rapid disappearance from our lives of face-to-face conviviality. It would mean trying to build societies that engender happier human beings, instead of doing what technoscientific capitalism urges us, increasingly, to do: design human beings that fit into the society which has emerged on its watch. And so cosmetic surgery and body modification are merely the early manifestations of a trend that is likely to become pervasive, as various types of 'wetware'

are developed, and CRISPR/Cas9 gene-editing technologies become ever more general in their application.

In the next three chapters, I'll look at some of the revolutionary technologies that are bringing about, or may soon bring about, this attitude of Promethean striving, while also challenging the view of humanity on which those interventions are based. I will do so not from a religious position, or from a position of body–soul dualism, but from a view of humans as essentially social, creative and intentional creatures, whose basic needs change so slowly over time that it is as well to regard them as permanent. As technoscientific capitalism continues to reorder our lives according to its own priorities of progress and individual (or consumer) choice, we need, I think, to evolve a new 'holism' that can challenge the atomistic vision of both human society and human being inscribed in its ideology. Retro-futurist prints and science fictions to one side, the City of the Future will be a monstrous place indeed if its technologies merely serve our desires while overlooking our deepest needs.

5

OFF-TARGET EFFECTS

Genetics and the Human Future

He Jiankui doesn't look like a mad scientist. Softly spoken and in a neat blue shirt, the young biophysicist is not surrounded by conical flasks full of bubbling green liquid, twitching cadavers or brains in bell jars. He stands in a well-lit lab, clasping and unclasping his hands as he talks, in a gesture of compassionate reassurance. He could be giving a pep talk to a group of new university students, or a medical update on your ailing grandmother. Certainly there is little in his demeanour that would suggest he has just moved biotechnology into an extraordinary (some might say monstrous) new era.

He Jiankui, China's 'Dr Frankenstein', explaining to the world why he modified the genome of unborn twins Lulu and Nana

The short YouTube video in which He appears was put together hastily, in response to a scoop in *MIT Technology Review*. Published in November 2018, this article claimed that He and his team had been modifying human embryos and transferring them to women's uteruses in an effort to create gene-edited babies – something the scientific community had long considered beyond the pale, at least until much more is known about the complex relationship between human genes and particular traits and predispositions. Not only did He confirm that report, he also revealed that gene-edited twins, 'Lulu' and 'Nana', had been born in October, to an HIV-positive father, 'Mark', and an HIV-negative mother, 'Grace'. Working out of the Southern University of Science and Technology in Shenzhen, China, He and his team had combined IVF techniques with advances in gene-editing technology to encode resistance to HIV into the DNA of the two girls. This 'gene surgery' had gone well, claimed He: 'The girls are safe and healthy as any other babies.' Since Lulu and Nana were delivered in secret, the world had to take his word for it.

It was the health benefits of this treatment that He chose to emphasise, but few were impressed with this line of reasoning. After all, we *know* how to prevent HIV – by practising safe sex, principally – and now have effective drugs to treat it. Moreover, in modifying the twins' germline cells (their reproductive cells), He's team had leapt ahead of scientists working to develop genetic therapies aimed at modifying somatic cells (the cells responsible for everything in the body *except* reproduction). Most scientists make a clear distinction between therapies targeting these two types of cells, aware that alterations to the former could lead eventually to changes in the gene pool. By targeting the germline cells in a fertilised egg, He opened the door to precisely that prospect.

He's actions also raised the possibility that modern genomics could be used for *enhancement*, as opposed to, or as well as, the treatment of disease – that it could be used, in other words, to confer an advantage on particular children by selecting for desirable traits. In his video, He denounces such applications. But his actions reveal just how grey the area between essential and non-essential treatments is. He's actions were not curative, after all: they were aimed at encoding resistance to a disease that the twins neither had nor were likely to develop. In what sense then was He's intervention medically justifiable?

That question was given added urgency when it emerged that the changes made to the twins' genome may also have affected their intelligence. By deleting the gene known as CCR5, He's team had effectively closed the gateway through which the HIV virus enters a cell. But this same alteration has also been demonstrated to improve cognitive function in mice, and to assist human brain recovery after stroke. Whether He's intervention has had a comparable effect on the twins, the world will have to wait and see. But should it turn out that Lulu and Nana have higher-than-average intelligence, the prospect of 'designer babies' will surely move a little closer. Procedures such as prenatal screening, where foetuses are checked for specific birth defects, and preimplantation genetic diagnosis, where embryos are genetically profiled prior to being transferred to a uterus, are already fraught with ethical issues. The ability to actually *edit* the genome – to 'turn on' or 'switch off' particular genes and select for specific characteristics – adds a whole new set of questions to the mix.

He's actions, then, were rather more momentous than his sober presentation would suggest. And yet, for all his notoriety, and for all the calumny heaped upon him in the weeks and months after his appearance on YouTube, there is a sense in which his intervention was in keeping with the spirit of the age – with the speed and direction of biotechnology, and of the technosciences more generally. It suited the Chinese authorities to paint He as a rogue actor or a maverick; and it is true that he had either ignored or subverted a number of ethical practices and standards. But as far as the broader issue goes, his actions, and the justifications he gave for them, sit solidly within the trajectory of modern biotech.

That trajectory has taken our species an incredibly long way in an incredibly short time. Farmers have known for millennia that breeding animals with certain traits will result in particular characteristics. But it was only in the nineteenth century, when our knowledge of the cell began to increase, that scientists really began to converge on the precise mechanism by which biological information is passed from a parent animal to its offspring. Today we understand not only the mechanism, but also how to modify it: having 'cracked the code' of life on Earth, we can reengineer its flora and fauna, combining traits from different species, cloning animals from a single cell, and controlling insects and invasive rodents by modification of their reproductive cells. The 1950s discovery of the structure of DNA by Francis Crick and James Watson (with more than a little help from Rosalind Franklin, who is often written out of the story) led to speculation that the scientists of the future would be accorded a priestly role. The contemporary biotechnologist is more like a minor deity than a priest – a status revealed in the accusation that this or that scientist is 'playing God', as well as in the religious metaphors employed by historians of modern genetics: 'the book of life'; 'the eighth day of creation'. As we've seen, the overall thrust of the technosciences has been towards a fundamental rebalancing of science and technology, where the practical and transformative priorities of the latter have come to dominate the former. Nowhere is that process more apparent than in the field of modern biotechnology.

It is to this process that we need to turn our attention, not because our knowledge of the human genome is insufficiently sophisticated to warrant such radical interventions, but because those radical interventions are set to rewrite – are already rewriting – the terms on which we engage with the world: with nature, and with the human community that determines our selfhood. Being privy to the book of life is one thing; being able to edit it is another entirely, and we need to consider the consequences that may occur should genetic engineering become more mainstream. When geneticists talk of 'off-target-effects', they are referring to the unwanted mutations that occur as a result of biotechnical interventions. But we need to attend to the *cultural* off-target effects of such interventions, as much as the purely medical ones. How are these radically new affordances changing us at the social level, and how are they likely to change us in the future, as our ideas of human flourishing change, and the market continues on its merry way, monetising and commodifying everything? These are not questions any scientist can answer. They are questions we must answer for ourselves, on the basis of what kind of creatures we are, and what we believe we owe to each other and to future generations.

The Eighth Day of Creation

In *Fully Automated Luxury Communism* (2018), the British writer Aaron Bastani puts a leftist spin on the Promethean view of technological development. While noting the revolutionary potential of recent genetic innovations, he insists that the latter are no different in kind from the selective breeding practices of the past: they are simply another great leap forward in humankind's mastery over unruly nature. Referring to the movie *Elysium* (2013), which depicts a world where biotechnologies are only available to the very rich, Bastani's only political concern is whether the new genetic technologies will be privately or socially owned. All other questions are beside the point, at least as far as he is concerned. As he puts it, with alarming insouciance: 'Before editing the human genome at scale such efforts should be subject to vigorous public debate. But how much difference is there between improving nutrition for health outcomes and optimising our biological programming? Not much.'[1]

The particular innovation that interests Bastani – and that interests pretty much everyone who follows developments in biotechnology – is the technique known as CRISPR/Cas9. Developed by French professor Emmanuelle Charpentier and American biochemist Jennifer Doudna in 2012, this technology relies on the ancient immune system found in a large number of bacteria, which deal with potential viral threats by incorporating strips of the virus's DNA into their own using an enzyme called Cas9. These newly formed sequences are the CRISPR, which the bacteria then use to produce ribonucleic acid (RNA) copies to recognise viral DNA and repel future attacks: the RNA, which translates genetic information into the proteins necessary for cellular processes, recognises the rogue DNA and destroys it instantaneously. The CRISPR/Cas9 gene-editing system mimics this naturally occurring process, but its practitioners 'program' the Cas9 protein with a strip of pre-designed RNA that guides it to the right part of the genome, where it cuts out the targeted DNA. This was the technique He and his team used to modify the genome of Lulu and Nana, and its applications are potentially vast. Already it has been used in agriculture to enhance the drought tolerance and nutritional value of crops, and to immunise industrial cultures against infection, while in medical research it has been used to create a range of genetically modified monkeys to serve as models for the study of diseases such as autism, cancer and muscular dystrophy. Other research promises higher-quality meats, disease-resistant livestock and the elimination of inherited diseases such as Huntington's, cystic fibrosis and sickle cell anaemia.[2]

As with Bastani's observations, the debate around CRISPR/Cas9 technology tends to swing between a recognition that here is something radically new, and a kind of technological fatalism – a feeling that, for all the power this emerging technique affords humanity, it is nevertheless continuous with the broader trajectory of control over nature that has characterised the human story. That mood is caught in Bastani's assertion that 'Genetic engineering is nothing new', which is true, in the sense that human beings have long understood how selective breeding can enhance or eliminate certain traits. But Bastani's assertion is also false in the sense that CRISPR/Cas9 technology allows us not merely to redirect processes that we recognise as part of nature, but to actively intervene in those processes. While selective breeding takes natural processes and redirects them towards human ends, genetic editing reconstitutes nature in ways that would never occur independently. Even compared to more recent techniques, CRISPR/Cas9 is a radical departure. Up until 2012, the cutting edge in genetic engineering was a process known as gene transfer, whereby whole genes were transferred to targeted genomes. By contrast, the new gene-editing technologies allow scientists to make changes *to the genes themselves*, cutting, copying and pasting strips of DNA as and where they deem it necessary.[3]

In this sense, the new genetics is the acme of a very different approach to nature than the one that has characterised human life for the better part of its development. Involving the application of engineering principles to biological organisms, 'synthetic biology' is solution-oriented: it is a thoroughly practical application of science to the natural world, up to and including human biology, emerging from an historically unique view of what it is permissible to do, and transforming our relationship to nature in the process. We will discuss later on whether we come to see ourselves, or our own bodies, as a 'standing reserve' – as mere matter to be manipulated, in line with the instrumental ethos of technoscientific capitalism. But for now let us note that despite the best efforts of the developers of CRISPR/Cas9 technology – Professor Doudna in particular – to instigate a 'vigorous' debate of the kind that Bastani envisages (before philosophically shrugging his shoulders), we are now in virgin territory.

One way to think of this 'solutional' view of science, and of the instrumental attitude towards nature that accompanies it, is to see it as an aspect of the informational worldview we examined in the previous chapter. For modern genetic engineering is the handmaiden to a view of life as composed of discrete parcels of information, in the same way that the computational view of the brain is the handmaiden to a view of consciousness as data. To some extent, this view of life flows naturally from the modern synthesis of Darwin's theory of evolution and Gregor Mendel's ideas on heredity: having discovered both the general processes according to which living organisms evolve (random mutation and natural selection) and the particular unit that carries information from one generation to the next (the gene), it became plausible to think of organic life as comprising graded continua, as opposed to fixed and permanent essences: life became, in part, the story of what's inside the things inside the things inside the organism.[4] But as science became increasingly central to the post-Enlightenment view of the world, this insight itself underwent a mutation. People began to think of life as *reducible* to those smaller elements, in a way that reinforced a view of human beings as essentially informational. In fact, *the whole world* became explicable in terms of those smaller elements, as the neurobiologist Steven Rose explains:

> The mode of thinking which has characterised the period of the rise of science from the seventeenth-century is a reductionist one.

Reductionism holds that to understand the world requires disassembling it into its component parts, and that these parts are in some way more fundamental than the wholes they compose. To understand societies, you study individuals; to understand individuals, you study their organs; for the organs, their cells; for the cells, their molecules; for the molecules, their atoms ... right down to the most 'fundamental' physical particles. Reductionism is committed to the claim that this is the scientific method, that ultimately the knowledge of the laws of motion of particles will enable us to understand the rise of capitalism, the nature of love, or even the winner of the next Derby.[5]

Rose is painting with broad strokes here, but it's clear that something like this tendency has emerged in the last three hundred years, and has become far more pronounced in recent decades. The question is whether the view is an accurate one, or the symptom of a mindset that regards the world as just so much information to be endlessly manipulated.

One factor driving the more relaxed view is how genetic technology has combined with information technology to underwrite the idea of life as an informational entity. The first silicon transistor was built in 1954, the year after Crick and Watson announced that they had identified the structure of DNA. And for the next half-century the ones and zeroes of computational technology and the As, Cs, Gs and Ts denoting DNA's four nucleotide bases were intertwined in a double helix of development, which culminated in 2003 with the completion of the Human Genome Project. Even before that announcement was made, some commentators had noticed that biotech was being made more permissible, and potentially more dangerous, due to the influence of the computer in scientific thinking. In *The Biotech Century* (1998), for example, Jeremy Rifkin noted how the prominence of computers in science led researchers to see nature in 'cyber' terms. Descriptions of genetic material as 'code', comparisons of consciousness to parallel processing, and even the word-processing metaphor favoured in explanations of CRISPR ('edit' ... 'cut' ... 'copy' ... 'paste') all serve to join information technology and genetic modification in the popular mind. The two informational perspectives reinforce each other, making further developments in genetic technologies more likely in the future.

Making those developments more likely still is the character of modern technoscience, which brings together the theoretical aspects of science and the practical ethos of technology, dissolving the distinction between 'pure' and 'applied' science that held good in previous centuries. Unlike during the Industrial Revolution, when factories were production-driven, technoscientific capitalism is driven by research and innovation, no more so than in biotechnology. The result is that new techniques and processes emerge with vertiginous frequency, outrunning our ability to think through their implications and subject them to proper scrutiny. At the time that Rifkin was writing his book, the biggest story in biotechnology was the cloning of Dolly the sheep in 1996 – an outcome broadly deemed impossible by scientists in the 1980s, and one that stunned a world still accustomed to thinking of cloning as the stuff of science fiction. (Steven Spielberg's movie *Jurassic Park* was released in 1993, just three years before Dolly was born.) A little over a decade later, it became possible not only to clone certain organisms, but to synthesise entirely new forms of life. In 2010, a team of scientists working under the biotechnologist Craig Venter was able to synthesise a complete bacterial chromosome and transfer it into a bacterial cell, which was then able to grow naturally. CRISPR/Cas9 was revealed two years later, and Lulu and Nana were born four years after that. As noble and sincere as Doudna's dream of an informed debate on these issues is, my sense is that the speed and direction of development makes such reflection difficult.

Reporting on Venter's remarkable breakthrough – the creation of the first artificial organism – *Wired* reached in its very first sentence for a computational metaphor: 'Man-made DNA', it wrote, 'has booted up a cell for the first time.'[6] But for others, the development was rather more significant than this breezy description seemed to suggest. One group of academic scientists described it as an 'atom-splitting moment' – a manifestation of 'hybrid technoscience' no less significant than the Trinity Test. 'It was a colossal achievement for biology,' they wrote, 'and its significance might well rank alongside the detonation of the first atomic bomb in terms of scientific advance.'[7] The question is: if Venter's creation of a synthetic bacterium was the Trinity Test, what then was He Jiankui's revelation that he had used the CRISPR/Cas9 technique to alter the genome of unborn twins? Without wanting to overcook the analogy, I think we can say that it was a lot more 'explosive' than the current muted debate would suggest, and that if we continue to focus principally on how these techniques transform biological *life*, as opposed to how they may transform human *living*, we will be placing future generations in great danger.

'And Yet It Moves'

One reason we focus less than we should on the social aspects of bio-technology is that debates about new discoveries and procedures leave no room for social and humanistic reflection. Instead, such debates almost always devolve into arguments between 'clearheaded' science on the one hand and religion and superstition on the other. The stem-cell research controversy is a case in point. As the cells from which all other cells develop – a sort of biological blank canvas – stem cells are especially prized in medical and reproductive research, but they are also very hard to come by. One rich source of stem cells is human embryos, and one rich source of human embryos is eggs that have been fertilised as a result of in vitro fertilisation but never implanted into a woman's uterus. Typically, the issue of whether it is justified to use embryo-derived stem cells in scientific research will turn on the question of whether that embryo constitutes a human being, with religious objectors arguing that it does and scientific researchers arguing that it doesn't. But this debate doesn't even begin to exhaust the issues such research throws up. As academic Kate Cregan has suggested, stem-cell research raises crucial questions about the long-term effects on the public consciousness of such manipulation of human material.[8] Even if we choose (as I choose) to believe that an embryo is not yet a human being, there is no doubt that using such embryos in research has the effect of objectifying material life, moving us further along the road to a new technoscientific reality in which Martin Heidegger's 'standing reserve' extends to human beings themselves. Such processes have profound effects on the way we understand the world and relate to our fellow human beings.

In some ways, indeed, the religious objections made to new techniques in genetic biotechnology let technoscience off the hook. In most technologically advanced societies, religion is associated with the hidebound and the conservative, to the extent that even many believers do not expect organised religion to influence social policy. When believers *do* express concern about developments in biotechnology, science is able (and more than willing) to identify itself with human progress and Enlightenment rationality. In excitable moments, it will invoke the trial of Galileo in 1633, when the ageing astronomer stood defiantly before the Roman Inquisition as it charged him with the crime of heresy for suggesting that the Earth moves round the Sun, rather than the other way around.[9] (According to one popular version of the story, the guilty verdict was met with the words *Eppur si muove* – 'and yet it moves' – a rebellious assertion of science's truth in the teeth of religious totalitarianism.) Such framing allows scientific commentators to channel the same technological fatalism that Bastani is channelling in his comments on CRISPR. The story of science is the story of progress, and anything that gets in its road is just reactionary or superstitious.[10]

Galileo being tried by the Inquisition in Rome in 1633. He is pushing away the Bible.

There is even a sense in which science and medicine are now treated as *inherently* ethical precisely because they are *not* associated with arguments from religious authority.[11] One can see this in the changing character of scientific archetypes in culture. The literature of the past is full of examples of scientists whose desire for knowledge is the source of hubris and catastrophe: from Mary Shelley's Frankenstein to Robert Louis Stevenson's Dr Hyde to H. G. Wells' Dr Moreau, novelists (not all of them hostile to science) have drawn a thick black line from the Promethean to the diabolical, channelling the cautionary legend of Faust in a generic indictment of scientific arrogance. Today, it is science, not religion, that tends to have prescriptive force. The scientist, far from selling his soul to the Devil in return for unlimited knowledge, is a model of sagacity and sanity – the one who reins in the belligerent general, or warns against too hasty a judgement regarding the octopoidal aliens parked in the skies above Montana. Those who say that science and technology are *themselves* a religion overstate the case. But that science now claims a kind of ownership over moral as well as technical questions is impossible to deny.[12] As the popularity of 'evangelical atheists' like Richard Dawkins and Sam Harris attests, many people now look to scientific endeavour and explanation for ethical guidance.

Is this a bad thing? In many ways, no. As compared to the manifold obfuscations and built-in prejudices of religious thinking, science has given human beings a clearer understanding of nature and of their (exceptional) place within it. Modern science is one of humanity's greatest achievements, and a source of the numinous as well as the factual. When, each morning, I turn on my computer, I am greeted by a photograph of an exploded star that hangs in space like a delicate jellyfish – an image from the Hubble telescope as inspiring as anything in the Book of Psalms. But in order to bring about a reflective relationship towards science and technology, we need to distinguish between the scientific *method* and science as a *social process*, and to keep that distinction front of mind. For while nuclear science may be able to tell us *how* to split a plutonium atom, it cannot tell us *if* we should do so, any more than a teapot can tell its owner when to have a cup of Darjeeling. Even leaving aside the fact that the story of biological science is *not* a story of unimpeded progress – as the history of eugenics shows – we are still left with the stubborn reality that science is a human pursuit with no inevitable trajectory of its own. As we face the prospect of a 'soft' eugenics in the form of CRISPR/ Cas9 technology, confusion on this basic point could have far-reaching and devastating consequences.

To put it another way, scientific autonomy is, and always was, a myth – a myth that no doubt stemmed in part from the character of the scientific process, which earnest proselytisers for science in general have read across to its institutions.[13] But that this myth has survived into the era of technoscience – an era in which science, utility and the market are thoroughly interwoven – makes it far more dangerous than it was in the past, when scientists could still pursue their research in relatively disinterested fashion. Now more than ever, scientific research is a highly organised social activity dependent on various forms of buy-in and support. In the space of just two centuries, it has gone from a largely amateur field (pursued, more often than not, *in fields*) to a profession based in the laboratory and the university – institutions now thoroughly integrated into commercial research and industry.[14] At the same time, and not incidentally, the notion of scientific progress has often been deployed in order to mask this social and economic context. As Jasanoff puts it:

> [O]ne recurrent simplification that science, aided and abetted by the mass media, performs about itself is to rid its progress narrative of all material incentives and supports – whether the hands of the modest, invisible technicians who turn the idea into product, or the spin-offs and start-ups that have turned the modern university into a hotbed of business entrepreneurship, or the recent agendas funded by Silicon Valley's tech titans who want to buy knowledge in order to lengthen their own unimaginably privileged lifespans.[15]

In this way, the characterisation of science as a process of disinterested discovery disguises its social and economic underpinnings and keeps critical humanistic reflection at bay.

It's true there have been controversies in the field of biotechnology, most notably around genetically modified crops such as Bt corn (which produces its own insecticide) and Roundup Ready soybeans (which are resistant to certain weed-control herbicides). But in broad terms, the incorporation of science into technoscientific capitalism puts enormous pressure on the institutions of oversight and ethical reflection, which find it increasingly difficult to keep pace with changing circumstances.[16] The philosopher Michael J. Sandel has written of the 'moral vertigo' such rapid development can create, as the popular esteem in which science is held protects it against criticism. This dynamic even extends to the institutions of oversight themselves. The famous Asilomar Conference, for example, convened in 1975 in the wake of the discovery that recombinant DNA could be used (in principle) to engineer cells, was heavily skewed towards questions of safety, in a way that left precious little space for more general moral and social deliberation.[17] As such, it effectively baked in the assumption of biotechnology's inevitability, and, by extension, the inevitability of a biotechnological industry – assumptions that have been reproduced in ethical discussions around biotech ever since.

In their introduction to *Human Flourishing in an Age of Gene Editing* (2019), Erik Parens and Josephine Johnston acknowledge the difficulty of speculating on the social (as opposed to medical) problems that may flow from existing and emerging technologies: 'saying what we mean when we say we're worried about nonphysical harms ... is much harder than saying what we mean when we say we're worried about physical harms'. But of course it is precisely *because* those discussions are so difficult to have that we need to have them, as a matter of moral and political urgency. The difficulty here is not incidental: it derives from the fact that when we talk about these issues from a moral and humanistic perspective, we are talking about *the kinds of creatures we are*, and what our human limits might be. We are talking about our humanity, and about the opening up of a liminal space in which uncanny, even monstrous, versions of ourselves appear to be moving about in the shadows.

A New Eugenics?

We only need to look to history to see what happens when biologism (the interpretation of life purely through the lens of biology) becomes the intellectual fashion. Social Darwinism, for example, founded in the nineteenth century by the philosopher and scientist Herbert Spencer, has been an unmitigated catastrophe for humankind. Unlike Darwin, who believed that 'fitness' came about through natural selection of small hereditary variations that affected an organism's ability to survive, Spencer believed that it came about through the inheritance of acquired characteristics: it was individual organisms striving to come to grips with their environments and struggling against their competitors that drove evolution. For Spencer and his acolytes, this process of 'survival of the fittest' was observable in human affairs, where inequalities of wealth and status were taken to reflect *biological* inequalities – an idea that proved popular with laissez-faire economists and colonial administrators eager to give their operations the stamp of scientific respectability.

Not that everyone who adopted the ideas of Social Darwinism was politically reactionary. Many of them were social progressives motivated by ideals of equality and the alleviation of poverty. It is vital to remember, in any discussion of bioethics, that not only Social Darwinism but also the related field of eugenics had far more currency among intellectuals than their association with the racist genocides of the twentieth century might suggest. Eugenic ideas were a central feature of Progressive Era politics in the United States, while liberal intellectuals such as H. G. Wells believed strongly in the use of eugenic science to effect an 'improvement of the human stock'.[18] Nor did Social Darwinism die with the Armenians and the European Jews. The social and economic policies enacted by the right-wing governments of Ronald Reagan and Margaret Thatcher – policies that stressed economic competition and downplayed the benefits of cooperation – descended from this common ancestor (albeit with some modification). Even the concept of meritocracy possesses clear affinities with it: the idea that people are born with different abilities, which are expressed at the level of income, success and influence, still suffuses politics and culture in liberal democracies as a sort of political common sense.

A pro-eugenics poster from the 1930s

Meanwhile, the notion that social arrangements have a basis in biology continues to command intellectual respectability via evolutionary psychology, which attempts to understand the human mind in terms of evolutionary processes. Hugely popular in the 1990s, this patchy discipline proved influential on some libertarian commentators called on to explain why market mechanisms haven't led to more social mobility. One common move was to claim that those on the bottom of the social pile have lower than average intelligence, and that those at the top are there because of their intellectual prowess. This was the argument Richard Herrnstein and Charles Murray made in their controversial book *The Bell Curve* (1994), which argued that the inferior socioeconomic position of black Americans could be partly explained by a finding that their scores on IQ tests were a fairly steady 15 points below the average non-black test-taker. (Since the authors claimed that intelligence was around 60 per cent heritable, this finding led many readers to conclude that a genetic element was at work in that differential – a conclusion not explicitly drawn by the authors but consistent with their overall argument, and with the political 'tone' of their supporters.) Even today, such ideas influence both conservative and libertarian thinkers. The psychologist Jordan Peterson, for example – for a time the darling of the alt-right trolls – derives a defence of social hierarchies from a comparison of human beings and lobsters. For Peterson, the fact that serotonin increases in a lobster's brain the higher up the social hierarchy it climbs is evidence that we, too, are hardwired to seek power and influence. As he puts it in *12 Rules for Life* (2018), his foray into the self-help genre: 'It's winner-take-all in the lobster world, just as it is in human societies, where the top 1 percent have as much loot as the bottom 50 percent – and where the richest eighty-five people have as much as the bottom three and a half billion.'[19]

Not all evolutionary psychology is as dumb as this. But much of it seems to be based on a fallacy, in that it takes the insight that human beings possess certain natural capacities (which manifest as particular traits) and extrapolates that *all* human traits must have a genetic cause. These thinkers will tend to look at, say, the human habit of worshipping gods, or playing sports, or wearing perfume, and assume that these highly complex behaviours must have an evolutionary explanation. Psychology is constructed as an extrusion of nature, in an instance of what the philosopher and neuroscientist Raymond Tallis calls 'Darwinitis': a 'pathological' version of Darwinism that claims to explain not only how human beings evolved but everything about what it means to *be* a human being, from human behaviour to human institutions.[20]

The arguments of evolutionary psychologists may seem a long way away from modern biotechnology. But the point is that such biologistic approaches to matters of social and psychological concern are always prone to treat human beings as individuals first and social beings second. In this sense, they draw on the same mechanistic vision of humanity discussed earlier. As the late Australian academic Anthony Smith put it in 1993, in an article on the Human Genome Project, 'The biological determinist (mechanistic) perspective embodies a significant component of individuation – the concentration of attention on the individual, the divorce of the individual from his/her socio-political environment, and the down-playing of the role of the environment in the individual's physical, emotional and social well-being.'[21] Once again, and as we have seen many times, the mechanistic view of life underwrites a view of human beings as self-directed individuals, not members of society.

This is why claims to scientific disinterestedness need to be treated with the utmost caution. For while most individual biotechnologists are no doubt good and honest people, the intellectual settings around medical science in general and biotech in particular tend to reproduce a view of society as a collection of individuals, as opposed to a view of the individual as shaped by society. When a new drug or medical procedure is minted, the mainstream media will invariably focus on the unwell or at-risk individuals who stand to gain from the development, as opposed to the broader ethical concerns. There is nothing dishonest about this: it is harder, as Parens and Johnston say, to talk about nonphysical harms than it is to talk about physical ones. But there are many more ways in which things can go wrong than mistakes or unintended consequences at the level of the individual patient.

In other words, the long-term consequences – or off-target effects – of new biotechnologies aren't always open to measurement. Yet the contemporary emphasis on choice and individual rights is so pronounced that we tend to forget these deeper social and ethical issues. To oppose a new procedure for reasons other than precautionary medical ones would be to oppose a liberal bias in favour of moral autonomy and choice.

To make choice the overriding principle is to adopt the language of neoliberalism, and thus to move a little closer to the version of reality favoured by those who regard society as an aggregation of individuals rather than a tightly woven fabric of deep (if often obscure) attachments. It takes us closer to the idea that society is a marketplace, and that freedom is reducible to consumer choice – a vision of society in which the concept of autonomy can often serve as a handy cover for decisions taken in the interests of profit. Anyone who believes that the priorities of capital are not driving practice in the biotech sector simply hasn't been paying attention.

For those of a neoliberal persuasion, there is nothing inherently wrong with this. If a company decides to invest its capital into research on Alzheimer's or Parkinson's disease, so much the better for society. Nor is there anything wrong, in principle, with letting people decide for themselves what products to buy or services to use. As long as there is ethical oversight, and no one is harmed or obviously exploited – as they are in the so-called red market in organs, and in parts of the surrogacy industry – then consumers should be free to make their own choices. And the more choice there is, the better for the consumer.[22]

Well, the choices are certainly multiplying, and are likely to proliferate into the future. In vitro fertilisation, combined with increased genetic knowledge, means doctors regularly screen embryos before transferring them to uteruses – a practice that opens up the prospect of selection for cosmetic traits, as opposed to viability and inherited genetic disorders. Already there are companies that offer customers genetic profiles of their embryos, giving prospective parents 'polygenic risk scores' (the number of genetic variants an individual has, linking to their heritable risk of developing particular diseases), as well as information on traits such as sex, eye colour and hair colour.[23] Some countries, such as the United States, allow sex selection for reasons beyond the medical, including for family balancing. And while, at the moment, it appears that most doctors and scientists are against such practices, it is often unclear on what basis those communities object to them. Comments to the effect that life is not a commodity, or should not be regarded as such, are well taken; but without any anchoring notion of the 'sacred' – of what is special about human beings, and how the notion of 'designer babies' (or 'positive eugenics') impinges upon that specialness – the market is likely to make further inroads. We are some distance from the genetic 'supermarket' philosopher Robert Nozick advocated for in *Anarchy, State and Utopia* (1974), where parents are free to choose attributes of their children without state interference. But it's surely not in serious doubt that reproductive medicine is moving in that broad direction.

And with all respect to the medical profession, I don't think it is in the best position – institutionally or intellectually – to arrest this.

Referring to the distinction that is often made between genuine diseases and 'suboptimal' traits that may affect one's chances in life (shortness, ugliness, having two X chromosomes), a wag once remarked that a healthy person is someone who has been insufficiently examined. As our knowledge of human genetics increases, we will need to evolve a much firmer sense of what we take the parameters of genetic engineering to be. Where does 'health' stop and 'beauty' begin, and the curative shade into cosmetic enhancement? Often, the objection to soft eugenics appears to be based on little more than the historical echoes of the noun in that phrase; but with utilitarian bioethicists lining up behind the likes of James Watson to say that it would be remiss of humanity *not* to engineer the human genome to eliminate, say, the very stupid, we will need a better response than mere unease. We will need to say where that unease originates, and why it is worth interrogating.[24]

Cogs and Sprockets

Historically, many of the more speculative objections to genetic engineering, cloning and so on have focused on the issue of class. In Aldous Huxley's *Brave New World* (1932), for example, the future World State has abolished parenting with a view to creating human babies that can be neatly fitted into one of five castes. Embryos are produced in hatcheries and passed along a conveyor belt in a grim pastiche of Fordist production. A similarly dystopian conceit is at work in Andrew Niccol's movie *Gattaca* (1997), which imagines a world in which genetic engineering underwrites social hierarchies. In such science-fictional scenarios, the evils of designer humans are likely to follow, and increase, the evils of society. Inequalities of power, status and wealth are written down in DNA.

As fantastical as those scenarios are, the points they raise are sound. In the event that genetic engineering becomes widespread in stratified societies characterised by inequality and discrimination, it will tend to reproduce that pattern. Indigenous communities that have been on the end of attempts at genetic 'improvement' in the past tend to be more wary than others about the claims of contemporary biotech, while the developing countries of the Global South might well object that genetic research in the countries of the Global North tends to focus on the kinds of diseases that richer people are likely to develop: diseases of old age, such as Alzheimer's, for example.[25] This is one reason radical commentators like Aaron Bastani are so keen to emphasise the importance of collective ownership. So transformative are the new techniques – and so rapidly will they reduce in cost, given their simplicity – that it is necessary to keep them in the public sphere, away from the system of patents and profits. For Bastani and his fellow 'accelerationists' (who seek to intensify growth and technological change, with a view to taking it over in future in the name of the working class), there is nothing implicitly wrong with the tech. The key question is which economic class is standing at the plunger end of the needle.

This call for a scientific commons is widely shared by utilitarian philosophers, whose habit, as we saw in Chapter 3, is to introduce into moral questions a calculus based on Bentham's principle of the 'greatest happiness of the greatest number'. Both Julian Savulescu and Peter Singer, for example, have no in-principle objection to gene editing, even for enhancement purposes, but are keen to avoid the distributive injustices that would result from an unregulated market in it. However, it is clear from the negative responses that these utilitarians tend to provoke when they enter into these fraught deliberations that moral reservations about genetic engineering go far beyond questions of ownership and access. On what basis would a utilitarian object to the creation of a human being (or even something close to a human being) for purely *utilitarian* purposes – a non-sentient 'saviour baby', perhaps, whose organs could be harvested in order to save a dying child? The scenario is fantastical, yes, but it is no more offensive than some of the conclusions reached by utilitarians themselves. In *Practical Ethics* (1980), for example, Singer argued that since a human infant at three weeks old is no more sentient or sensitive to pain than a foetus in the womb, to euthanise it is morally permissible in certain (vaguely defined) circumstances.[26] Needless to say, many people regarded that suggestion as monstrous, and not because such procedures would only be available to those with the money to pay for them.

Singer's answer might be that they objected to it emotionally, or on religious or quasi-religious grounds, and so are thinking irrationally. This is a common move among those who advocate controversial developments in the biotech space – to accuse their critics of invoking 'the yuk factor', otherwise known as the 'wisdom of repugnance' or (within logic) 'the appeal to disgust'. But while feelings of disgust should not in themselves be taken as evidence of the rightness or wrongness of a particular action (and true too that such feelings are often connected to some of our nastiest prejudices – racism, misogyny, homophobia), they are nevertheless an important point of departure in many ethical questions. As the philosopher Mary Midgley put it: 'Feeling is an essential part of our moral life, though of course not the whole of it. Heart and mind are not enemies or alternative tools. They are complementary aspects of a single process.'[27] If we believe (as Midgley did) that there is such a thing as a human nature, and that both emotion and the capacity for reason are intermingled aspects of that nature, then we need to ask what our disgust responses are telling us, while also noting the emotional element may underlie 'purely rational' positions.

Midgley is not the only philosopher to take emotional or intuitive responses seriously. In his lectures on biotechnology, Michael Sandel will invite the audience to respond in thumbs-up/thumbs-down fashion to ethical scenarios related to the use of new biotechnologies, before digging down, anthropologically and morally, into what those responses might reveal. Concerned not to veer off into wild speculation of the *Brave New World* variety, Sandel stays close to ethical issues that are either in play or on the horizon, such as the use of human-growth hormone for very short children, or whether it is ethical to use preimplantation genetic diagnosis or 'sperm sorting' to choose the sex of one's children. He understands that these apparently minor issues are the first to make visible the moral and cultural dilemmas that will occur when genetic engineering techniques such as CRISPR/Cas9 become more available than they are today. Should treatments aimed at promoting muscle growth in people with muscular dystrophy, or at arresting the deterioration of memory in people with some varieties of dementia, be marketed to the general consumer as bodybuilding aids or cognitive enhancers? Should prospective parents be able to select for the abilities and physical appearance of their children? Such questions are no different in principle from the ones Sandel tends to raise in his talks.

For Sandel, the 'fairness' or 'equality' objection to enhancement is limited – a point he elucidates with the theoretical scenario of a genetically augmented athlete. Most people, I imagine, would accept that an athlete who has undergone genetic therapy in an effort to enhance their performance has morally disgraced themselves in some way, as well as polluted the sport in which they are a participant. But since it is always the case that some athletes are genetically better endowed than others, our objection cannot be founded on fairness. 'From the standpoint of fairness,' Sandel suggests, 'enhanced genetic differences would be no worse than natural ones, assuming they were safe and made available to all.'[28] It follows that if genetic enhancement in sports is morally objectionable, it must be for reasons *other* than fairness. Our intuition must have some other basis.

Again, this is a reasonably trivial scenario, certainly compared to the kinds of issues that are likely to arise in the future. But it is a revealing one, in that it speaks to something fundamental in our sense of what it means to be human. An athlete is celebrated not merely for her ambition, or even for the effort she puts into her training, but rather for her *excellence*, and this excellence depends on the use she makes of her *natural* abilities. Sandel:

> The real problem with genetically altered athletes is that they corrupt athletic competition as a human activity that honours the cultivation and display of natural talents. From this standpoint, enhancement can be seen as the ultimate expression of the ethic of effort and wilfulness – a kind of high-tech striving. The ethic of wilfulness and the biotechnological powers it now enlists are arrayed against the claims of giftedness.[29]

It is, I think, this notion of giftedness that speaks to us at the level of our essential humanity. And while this may seem to fly in the face of the modern belief in reward for effort and equality of opportunity, it substitutes for those rather flimsy concepts a much deeper idea of human connection – one based on the fact that we arrive unbidden, not as objects of conscious design, but as beings with dignity, ends in ourselves. On this view, *all* forms of eugenics, positive as well as genocidal, are fundamentally offensive. It may even be that human solidarity depends on something like this sense that we are all equal in coming from nature, and that attempts to reengineer ourselves, to turn ourselves into our own pet projects, will prove corrosive of this deeper equality. When Singer asserts that a three-week-old infant can be euthanised because it feels no pain, he trespasses on this solidarity, because we recognise that its purest form is the love we feel towards our children: a love that is *unconditional*, and so powerful that it spreads beyond the parameters of the individual family to encompass children as a group – a group we regard as uniquely precious for no other reason than that they are children.

But that's the problem with utilitarianism, and with the system of technoscientific capitalism whose logic it tends to reproduce *regardless* of the stated political allegiances of its adherents: it introduces calculation into areas of life that we value precisely because of their incalculable quality.[30]

In a debate with Singer on bioethics, Sandel uses the word 'monstrous' to describe a theoretical scenario in which a chicken has been genetically modified in order to remove the instinct to roam. Midgely, too, is not afraid to use the word in her thoughts on the subject of bio-engineering. In fact, she goes further, and asks us to consider what wisdom this idea of the monster might bring to the debate on biotechnology, concluding that it 'centres on the concept of a species' – of where a species' parameters lie, and what the consequences of expanding (or obliterating) them might be. In other words, the notion of the monster recalls us to the question of what makes us what we are, and warns us to respect the answer as something, finally, beyond our control. It contains an injunction against hubris, and against the category mistake implicit in the very notion of bio-*engineering* itself. Of this 'simple analogy with machines', Midgley writes:

> Cogs and sprockets can in principle be moved from one machine to another since they are themselves fairly simple artefacts, and both they and the machines they work in are more or less fully understood by their designers. Those who use this analogy seem to be claiming that we have a similar understanding of the plants and animals into which we might put new components. But we did not design those plants and animals. This is perhaps a rather important difference.[31]

A rather important difference, yes – but one that the Prometheans who regard the reconstitution of nature as the working-out of human freedom are apt to miss, and to go on missing. 'Society will decide what to do next,' He Jiankui told the Associated Press, shortly after the story of Lulu and Nana broke. If there is a gene for disingenuousness, 'China's Dr Frankenstein' has inherited it.

6
PROJECT CYBORG

Escaping the Body, Escaping Ourselves

'You've been a good and faithful servant, Severus, but only I can live forever,' hisses the evil wizard Voldemort in *Harry Potter and the Deathly Hallows*, before slashing Professor Snape's throat with a spell. Through this pitiless act, the Dark Lord hopes to win the allegiance of the Elder Wand, one of a trio of magical objects that together afford mastery over death. Legend has it that the wand gives its loyalty to the wizard who wins it from its previous owner. Since it was Snape who murdered its former owner – Hogwarts' headmaster Albus Dumbledore – command of it will now pass to Voldemort. Or so the Heir of Slytherin believes, as he orders his trusty snake Nagini to finish the Potions professor off ...

Like many things in J. K. Rowling's hugely successful series, this scene has deep roots in the literature of the past, and here the roots go all the way down into the origins of literature itself. Composed in ancient Mesopotamia between three and four thousand years ago, the Babylonian *Epic of Gilgamesh* is the story of a Sumerian king who embarks on a quest to find the secret of eternal life, but who comes to accept his mortality as the natural and proper will of the gods. Not to accept it would be a form of hubris, and it is this theme that is taken up in so many of the tales and legends that bear on this question of eternal life, with their elixirs, fountains, grails and pacts. In some traditions, the desire for immortality involves sacrifice, magic or some unnatural act, as when the vampire must drink the blood of innocents. In others, immortality is experienced as a curse. The desiccated Struldbruggs in Jonathan Swift's *Gulliver's Travels* and the hedonistic protagonist in Oscar

Wilde's *The Picture of Dorian Gray* (1891) are in this latter tradition, as are the 'undead' pirates in *Pirates of the Caribbean* (2003) – cutthroats doomed to wander the oceans in search of the final piece of treasure that will return them to their mortal form. Thus are evil and eternal life – immorality and immortality – connected in both high and low culture, in a way that may be telling us something interesting and important about ourselves.

Sigmund Freud, for one, was convinced that the desire for immortality needs to be repressed if human beings are to develop as they should. For Freud, as for his colleague Otto Rank, this desire for immortality is evident in the child's creation of 'doubles' – projections of the self that shore up the ego, much as an ancient king might sustain a sense of his own magnificence by populating a city with statues of himself. But as a child matures, it learns to repress this narcissistic desire to cheat death, and the double becomes a thing to be feared – not a bulwark against death but a reminder of it.[1] Doppelgangers and evil twins are versions of such 'monstrous' doubles, whose monstrosity, for Freud, is a variety of the uncanny. Maladjustment, immortality and monstrousness are connected at a deep level. The urge to cheat death – to 'live forever' – is in some way implicitly wrong or wicked.

Dante Gabriel Rossetti's *How They Met Themselves* (c. 1860–64) depicts a couple in medieval dress meeting their doppelgangers in a dark forest

And yet, and yet ... in the sunny climes of Silicon Valley, a ghoulish assortment of contemporary Voldemorts dream of doing precisely that. Some are health freaks, pure and simple, obsessed with fasting and superfoods, ketogenic diets and natural supplements. Others inject themselves with hormones, wear Oura smart rings to measure their sleep patterns, insert glucose monitors under their skin, or pay several thousand dollars a pop for a litre of adolescent plasma in a bid to stave off heart disease, Alzheimer's, memory loss and multiple sclerosis.[2] Still others fund serious ('serious') research into new genetic biotechnologies that will allow them to live, if not forever, then for very much longer than is currently feasible. In 2013, Google announced its acquisition of Calico, a company that aims to combat ageing and diseases of ageing in a bid to 'solve death', while PayPal co-founder Peter Theil continues to invest in the SENS Research Foundation set up by the British gerontologist and mathematician Aubrey de Grey, whose ambition to develop 'strategies for engineered negligible senescence' (hence 'SENS') has long been the toast of would-be immortals. Sporting a frizzy, Rasputin-like beard, de Grey claims to have identified the 'seven types of ageing damage' – including cell loss, mitochondrial mutations and intracellular waste products – as well as the biomedical interventions needed to arrest degeneration in each case.[3] So confident is he that he is on the right track, he has claimed that the first person to live to 1000 is probably alive today, and may even be fifty or sixty already. To the argument that such longevity, if made available to the general population, would result in catastrophic overpopulation, de Grey counters that delayed onset of the menopause would mean that women could wait longer to have children, thus *slowing* the rate of population growth overall and buying ourselves a few centuries in which to figure out how to establish space colonies. And yes, he does appear to be serious.

The general term for such interventions is 'biohacking', and it seems clear that, for many in Silicon Valley and its technological hinterlands, immortality, or extreme longevity, represents the ultimate 'hack', of a piece with the latest developments in virtual reality, robotics and biotech. The desire to cheat death is related to the machine-like view of humanity that dominates in such high-tech circles. For all their obvious magical thinking, the New Immortals are in thrall to the idea of the future as set out in transhumanism: a future in which humanity merges *physically* with smart technologies, whether in the form of nanobots that deliver drugs with incredible accuracy, biochemical preparations that enhance our cognitive capacities, or the brain–computer interfaces dreamt of by Elon Musk and his equivalents. They dream of a cyborg future, in other words, in which organic and non-organic systems meld into a single being, no less human for having been so melded. This is the Kurzweil philosophy: the belief that it is in our stars to outsoar our creatural form, becoming something *better* in the process. It is an ethos of *enhancement*.

It is easy to laugh at the New Immortals. In fact, it should be obligatory. The do-it-yourself biologists who inject themselves with CRISPR DNA, and argue against any regulation of new gene-editing technologies, are feebleminded libertarians, absurdly enamoured of their own example, while entrepreneurs who spend thousands of dollars on teenage blood and oestrogen blockers are self-satirising narcissists.[4] But in mocking such ridiculous endeavours, we should bear in mind the broader processes of which they are the (morbid) symptoms, and consider the basis on which we presume to dismiss them as ridiculous. When the New Immortals assert that in a world of pacemakers, hip replacements, artificial limbs and cochlear implants, human beings are *already* cyborgs, they are making a legitimate point. Those who are alarmed by the trajectory, and enveloping ethos, of such innovation are therefore called upon to say in what sense they regard that trajectory as wrongheaded. One of the leading figures in AI, Marvin Minsky, saw the cyborg future as an inevitable step in humanity's evolution. Was he right? And if not, in what sense was he wrong?

In the previous chapter, I referred to the philosopher Michael Sandel's hypothetical case of a genetically augmented athlete, which Sandel uses to explain the concept of what he calls 'giftedness'. No doubt the same point could be made by swapping the genetically augmented athlete for one using some kind of performance-enhancing prosthesis. But the issues raised by such symbioses of human beings and technology go beyond the question of what *kind* of bodies we elect to build through biotechnologies to the question of whether the human body is something worth maintaining at all. To listen to many of the New Immortals, one would have to conclude that something like the transcendence described by Kurzweil and his devotees is desirable. My sense is that this predilection is only the most extreme expression of a general feeling at large in the world – that our freedom depends on escaping the body, or at least on bringing it into line with who we think we *really are*. The poor reception given to Zuckerberg's plans for a disembodied Metaverse was encouraging; but in certain respects, and as we've seen, the Metaverse is here already. Bad ideas trickle down, even if wealth doesn't.

At any rate, these are issues that affect not just the Wizards of Silicon Valley, but ordinary Muggles like you and me. As science fiction becomes science fact, we had better take them seriously.

Natural Born Cyborgs?

As an encapsulation of the circumstances in which humankind now finds itself, Christoph Niemann's cover illustration for the 2019 technology issue of *The New Yorker* is hard to beat. The comic-strip-style image consists of four panels, the first three of which show the same bespectacled figure operating three different kinds of computer: a 1970s-style mainframe system, a 1980s-style PC and a modern laptop. All three computers are plugged into the mains, and their screens emit a bright white light. In the final panel of the strip, however, it is the bespectacled figure who is plugged into the mains, and who radiates the light. The boffin has merged physically with the technology.

The illustration is entitled 'Evolution', but Niemann is being at least a little ironic. After all, the cover depends for its humour on the surprise of that final frame: as things stand, the majority of us don't quite accept that we are destined to physically merge with our MacBooks. And yet, for a certain kind of thinker, such a prospect can be slotted neatly into an evolutionary narrative – not in the sense that random mutation and natural selection will arrange for that outcome (fortunately, Silicon Valley tech bros do not produce more offspring than the rest of us), but in the sense that *Homo sapiens*, as a species dependent upon technology, *was a cyborg from the very beginning*. According to this view, technological innovation is simply an extension of human beings' 'true' nature. It is the working-out of the human *telos*.

This perspective has some basis in fact. As I noted in the introduction, the use of tools predates *Homo sapiens* by an estimated three million years, and helped determine the kind of creatures we are now. The ability to manipulate fire allowed our hominid ancestors to partially pre-digest their food through cooking, which reduced the amount of energy needed for chewing and digestion. In the long run, that meant smaller guts and larger, more sophisticated brains that could be turned to other practical problems, such as how to make better shelters or garments, or how to cooperate to hunt large beasts. This translated into yet more energy, which translated into larger brains, which translated into more technology, and so on and so forth, for hundreds of millennia. Human technology and human intelligence co-evolved in a spiral of mutual development. Technology made us what we are, and now we fill up our lives with technology.

It follows that there is nothing unnatural about technology. The ability to make fire or flour or a soufflé doesn't mean that human beings have broken their programming; it's not as if we left nature behind when we designed the stone axe or the coracle, or even the steam engine or the mobile phone. We are animals, but animals of a special sort. We are *Homo faber*: 'man the maker'. Not knowing one end of a power drill from another is no disqualification from this condition. When it comes to technology, we all have a competence. Everyman is a handyman.

But the idea that technology is therefore just an *extrusion* of our nature is based on a simplified view of humanity. For while biology changes over hundreds of millennia, according to the laws identified by Darwin, technology changes rapidly, according to the process we call 'culture'. Another way of putting this is to say that our species has two trajectories:

natural evolution and cultural development. The first proceeds according to a process of gene mutation and reproductive advantage; the second as a result of collective knowledge passed from one human being to another, and from one generation to the next. And as human beings get steadily better at controlling their environments and protecting the physically vulnerable from harm, the force of natural evolution is blunted. In Shakespeare's time, only one in three humans made it to the age of twenty-one. Today, nearly 99 per cent of them reach that milestone. Again, human beings are no less *natural* than they were in the Palaeolithic era. But while jellyfish, stoats and crocodiles rely principally on their DNA for instructions on how to navigate their environments, human beings have both their DNA *and* collective knowledge to guide them. The result is that we are both *within* nature and *in a relationship with it*. So far as we know, this makes us unique.

Yet while it is in our nature to have a culture, that doesn't mean they amount to the same thing, and this is the point that the New Immortals and the Kurzweilian transhumanists contrive to miss. Beginning from the observation that human beings are adapted to using tools, they conclude that technological innovation is itself a sort of force of nature. This attitude towards technological development is caught, albeit in an indirect way, in the increasingly popular Big History genre, which places the rise of human beings to planetary dominance in the broader story of the universe. Set out in books such as David Christian's *Origin Story* (2018) and Yuval Noah Harari's *Homo Deus* (2016), Big History claims that the universe in general, and human societies in particular, have advanced across a number of energy thresholds – from the Big Bang to the Industrial Revolution – towards ever greater complexity.[5] Though far from complacent about the state of the planet, it nevertheless regards innovation and development as inevitable, combining natural and human history in a way that effectively collapses the two, or folds the latter into the former. Human culture and history are acknowledged, but only as secondary considerations, since everything in the human past is in essence a biological process that is no more than a single episode in the wider biology of the planet.[6] Thus, the evolution of humanity is framed in a way that necessarily obscures – or discounts – much messy detail (the struggle for control over resources, colonialism, class, women's rights and so on).

We can see immediately how such an outlook would appeal to a certain Promethean mindset, and it comes as no surprise that Big History is especially popular in Silicon Valley and other centres of economic and technological power. Both Christian and Harari have been invited to appear at the annual meeting of the World Economic Forum in Davos in recent years, and Christian received US$10 million from Bill Gates to develop the Big History Project, an online course for high-school students.[7] But perhaps the most significant aspect of this blending of biology and technology is the way it implicitly opens the door to the cyborgisation of humanity. If human beings were cyborgs from the start, why not embrace our immersion in technology as inevitable and liberating?

In fact, I think we can discern the beginnings of this Big Historical approach to humanity in the concept of the cyborg itself. The word, a portmanteau of 'cybernetic organism', first appeared in a 1960 article by Manfred Clynes and Nathan Kline. Their subject was humanity and technology in the coming era of space exploration:

> Space travel challenges mankind not only technologically but also spiritually, in that it invites man to take an active part in his own biological evolution … In the past evolution brought about the altering of bodily functions to suit different environments. Starting as of now, it will be possible to achieve this to some degree *without alteration of heredity* by suitable biochemical, physiological, and electronic modifications of man's existing modus vivendi.[8]

For Clynes and Kline, human beings' evolutionary and technological development were poised to combine, but this was not something to be alarmed about. On the contrary, it would allow us to boldly go where no human being had gone before.

There's a problem with this picture, though, and in the passage above it is represented by the mild-mannered adjective 'suitable'. For it is humanity's ability to *ask* what is suitable – what is good, what is bad, what is progress, what is regress – that separates it from other species. For Kurzweil and the New Immortals, technological development is a *neutral* quantity, an expression of the human essence: even if not driven by natural selection, it is part of our broader evolution, which is always pointing to greater complexity, as shaped by the 'law of accelerating returns'. But such a fatalistic account leaves out the very human qualities – reflection, judgement, caution, wisdom – that shape our way of being in the world, and of living in community with others. The use of the word 'suitable' in the passage above assumes a particular destination. The question is, suitable for *whom*, and for *what*?

To put it in technological terms: whose instruments are we using here?

A Fitbit in Your Skull

If there is one class of technologies that speaks to the relationship between humanity and its nature, it is medical tech. By using tools, human beings are able to protect themselves from the harsher elements of their environment, and to remedy many of the injuries that befall them. In this sense, the Big Historians are right to note our partly cyborg nature. Drugs and dentures, pacemakers and prostheses, walking sticks and wheelchairs: these are the emblems of human ingenuity – the outward sign of *Homo sapiens'* remarkable creativity and resilience. One of the world's greatest physicists, Stephen Hawking, was confined to a wheelchair for much of his life and communicated through a speech-generating device – both an example and a symbol of this species resilience, and a reminder that a more reflective attitude to new and emerging technologies should in no sense lead us to underestimate the vast strides human beings have taken in the treatment of disease and infirmity.

But better health and technological innovation do not advance hand in hand, like old lovers. We may welcome organ transplantation as an innovation that preserves human life, but as the philosopher Jacques Ellul pointed out in an interview in 1992, one reason there are so many organs available for transplantation is that people die in road accidents, as a consequence of a technology – the internal combustion engine.[9] Nor is there always agreement on the question of what constitutes legitimate intervention. For some in the Deaf community, devices such as cochlear implants represent an unwarranted medicalisation of their condition, deriving from nineteenth-century debates about what is and isn't a 'normal' body.[10] Or take the case of American child Ashley X. This profoundly disabled six-year-old had her uterus and breast buds surgically removed, and was also given oestrogen injections that reduced her height by some thirty centimetres. These procedures were performed, in part, to make Ashley more transportable, leading critics to suggest that her parents and doctors had effectively redesigned her body in line with their own requirements.[11] That's a harsh judgement, but such cases recall us to the blurred line between medical intervention and enhancement, and to the interests of those who would blur it still further.

Certainly Silicon Valley–style fantasies of bodily reconstitution will often hitch a ride on the practices and priorities of medical research. Just as food companies use clever packaging to make their products look healthier than they are, so Big Data launders its ambitions for power and profit through the language of 'wellness'. The almost total incorporation of medical research and development into technoscientific capitalism makes such corporate 'healthwashing' a good deal easier: in subordinating science to utility, and utility to the profit motive, we ensure that the pressure is always on manufacturers to expand and develop a product's applications, as when pharmaceutical companies slightly modify their drugs in order to extend their patents in a process known as evergreening.[12] The so-called wellness industry is in part a consequence of this systemic pressure. Since the genuinely sick are always in the minority, new markets, and new maladies, need to be found. The result is that 'good' or 'better' health is an ever-expanding category in which notions such as 'potential' and 'performance' are always working to blur the distinction between health and enhancement, the curative and the cosmetic. When the New Immortals claim that old age itself is a disease of the body, they are channelling this blurring of categories, and the false equivalence to which it leads.[13]

The slide from health to enhancement is perhaps most clearly discernible in the case of brain–computer interfaces (BCIs), communications systems that link the brain to an external computer. Such systems, which can be invasive or non-invasive, have a range of medical applications, from transcranial magnetic stimulation for the treatment of severe depression, to the prevention of epileptic seizures, to deep-brain stimulation treatment for sufferers from Parkinson's disease. It is, however, the BCIs that allow people with severe disabilities to complete tasks using their minds alone that attract the most interest from the big end of town, as well as from the military – systems that (to take one example) allow fully paralysed individuals to spell out words using only their thoughts, which are communicated as audio signals via brain implants linked to a computer.[14] Elon Musk's Neuralink are working to develop such applications, and are open about their ultimate aim of developing BCIs for the able-bodied consumer. They state that they want to produce BCIs with a bidirectional capacity, linking human beings to computers in a way that extends their cognitive capacities and achieves 'symbiosis with artificial intelligence' – the very situation flagged in Niemann's illustration in *The New Yorker*. And if that sounds like a long way away from the harrowing experience of paralysis or Parkinson's, the SpaceX founder is ready with a bit of wellbeing-speak to set your mind at rest. The Neuralink, he says, will be 'like a Fitbit in your skull'.

Elon Musk with his Neuralink machine

Musk may not get his wish. Despite the progress Neuralink has made (on the back of many others' progress), the dream of human–computer symbiosis – of a cyborg future in which human intelligence and artificial intelligence combine – could founder on the stubborn fact that the organic brain is very different from a non-organic computer, as we saw in Chapter 4. Nonetheless, if it *does* transpire that human and non-human intelligence can be linked in the way that Musk envisages, the creature that emerges from the union might very well not be human at all. As the philosopher Slavoj Žižek has noted, such an outcome would fundamentally alter the relationship between our inner lives and external reality, effectively obliterating freedom in the process. As Žižek puts it: 'We are free in our thoughts precisely in so far as they are at a distance from reality ... Once our inner life is directly linked to reality, so that our thoughts have direct consequences in reality – or can be directly regulated by a machine that is part of reality, and are in this sense no longer "ours" – we effectively enter a post-human state.'[15]

Žižek is not suggesting that human beings will lose their essential freedom as a consequence of surveillance capitalism. No doubt that phenomenon will play its part in future technology; but the issue here relates to something more fundamental than individual privacy. It relates to what Žižek, following Marx, would call our 'species being' – our ability (unique among the animals) to reflect upon our situation, and to act on the basis of that reflection. I think that a Neuralink, or something roughly akin to it, would end up undermining the conditions for both human freedom *and* human flourishing, by obliterating psychic distance and physical proximity at a single stroke. The science-fiction dystopias that present the melding of human beings and machines as a social and spiritual catastrophe are here a better guide to the future – a more *reasonable* guide, at any rate – than Kurzweilian fantasies of human transcendence, with their Judeo-Christian overtones. Reportedly a science-fiction fan who took his inspiration for Neuralink from Iain Banks' *Culture* series, Musk appears to have missed this aspect of futuristic literature.

Whether Musk's Neuralink ever enters mass production, the broader idea of health and happiness that animates him, and his fellow entrepreneurs in the information economy, is leading us into danger nonetheless. For them, it is in the name of progress that we need to develop these technologies; but the idea of progress that dominates in such circles is so clearly ideological, a *projection* of the system to which those entrepreneurs are wedded, that it simply reproduces that system. 'Obviously we need to keep innovating,' the Titans of Silicon Valley declare, 'because innovation is *what we do*. Without it, we would never have left the cave.' But it is not enough to leave the cave. You need to know what you're leaving it *for*, and where you want to get to in the long run. As a system that seeks high private returns, rather than high social ones, capitalism is perhaps not the greatest guide.

It is here that the concept of individual choice is so useful to information capitalism. For in insisting on the principle that they are only giving consumers what they want, the curators of our cyborg future can reconcile their Prometheanism with the priorities of the 'free' market. Moreover, they can speak back to those, like Žižek, who would challenge them on the question of freedom. For what greater freedom could there be than the freedom to exercise (consumer) choice! Asked about the many ethical concerns raised by human-machine technologies, academic and data scientist Davide Valeriani told *The Guardian* that 'one way to overcome these ethical concerns is to let humans decide whether they want to use a BCI to augment their capabilities'.[16] But this is to ignore the ideological process by which such choices are framed and made. The particular notions of progress and freedom favoured by the Titans and the New Immortals are not limited to the San Francisco Bay Area. They run through government policy, the media and university syllabi. Like Musk's imagined BCI, they are wired into our consciousness.

The good news is that it's not impossible to break out of this ideological bind. Precisely because they are founded on self-interest, the notions of progress and freedom on offer from info-capitalism are open to challenge. The popular distrust of Big Data that has crystallised in recent years may be an encouraging sign. But should a more rounded idea of freedom become bound up with projects such as Neuralink, we could find that technological development takes a sharp turn in the direction of transhumanism. In fact,

I would argue that transhumanism has already received such an injection of energy, and that for an emerging generation of activists, technological transformations of the body hold out not merely the prospect of progress, but the promise of political and personal liberation. These activists tend to see themselves as sitting on the radical end of the spectrum, so this dynamic isn't always obvious, and in one way it counts as an irony that some of the most strenuous arguments for the cyborg future are to be found on the radical left. Then again, perhaps this is the logical endpoint of Langdon's 'technological somnambulism': a situation in which a radical cohort is preparing to sink its differences with the billionaires of Silicon Valley on the question of technological transformation.

An Invention of Recent Date

First published in 1984 and aimed largely at an academic readership, Donna Haraway's *A Cyborg Manifesto* is not, perhaps, the most popular text in the book clubs and reading groups of Silicon Valley. In fact, I'd be surprised if it even makes an appearance on the shelves in Zuckerberg's virtual lair, or in the library of Jeff Bezos's Kindle. Radical feminist disquisitions on the liberatory potential of the cyborg metaphor are, I imagine, surplus to requirements in the testosterone-charged world of info-tech. No doubt Sheryl Sandberg gets a look in, with her brand of corporate ('lean-in') feminism, which was forged in that very environment. But post-Foucauldian deconstructions of traditional feminism? Hmm, not so much.

In fact, it is probably fair to say that Haraway's book-length manifesto is not much read by anyone these days, or by anyone outside the fields of post-humanism and gender studies. But a book doesn't need to be a bestseller in order to be influential, and there is no doubt that *A Cyborg Manifesto* made a splash not only in academia, but beyond it, in the world of political activism that emerged from the ruins of the so-called New Left. In common with much of the writing that appeared under the headings of post-structuralism and postmodernism, *A Cyborg Manifesto* is dense, allusive and often incomprehensible. But its basic thesis was easy to grasp, and it struck an intellectual chord that continues to vibrate to this day.

'Manifesto' is a slight misnomer. The text is a philosophical essay that rethinks the nature–culture distinction, suggesting that modern human beings should not be seen as animals with any essential qualities but as hybrids of nature and machine: as cyborgs. According to Haraway, this cyborgisation comes about through the erasure of a number of boundaries. First, Charles Darwin's *On the Origin of Species* erased the boundary between humans and animals, undermining the notion of human exceptionalism. Second, the Industrial Revolution erased the boundary between humans and machines, beginning the process of mechanisation of every aspect of human life. And, thirdly, the miniaturisation of technology (or its disappearance into software) erased the boundary between the physical and non-physical, in a way that has led to ambiguity about where humans end and machines begin. So, while a philosopher might once have argued that human beings were distinct from other animals due to their capacity for reason, and distinct from the world of artefacts due to their organic (animal) nature, for Haraway there was only ambiguity as to whether humans are natural or artificial. All the old dualisms dissolve into a reality of which the cyborg is the primary inhabitant.

Though encrypted in academic jargon, *A Cyborg Manifesto* marks an important moment in radical thought in general, and in feminist thought in particular. By the 1980s, the great social movements of the 1960s and 1970s – second-wave feminism, Black nationalism and so on – had devolved into a narrow identity politics that was increasingly 'intersectional': as concerned with the struggles *between* different groups as it was with challenging capitalism. Using the figure of the cyborg as a metaphor, Haraway urged radicals to reject that politics, and the 'essentialist' theories that underpinned it, from the perspective of the 'chimeric' and the 'monstrous'. 'Cyborg unities are monstrous and illegitimate,' as she put it, in her sphinx-like way; 'in our present political circumstances, we could hardly hope for more potent myths for resistance and recoupling.'[17] For Haraway, the cyborg metaphor was the model for a new solidarity, based on something other than the search for what she called 'revolutionary subjects' – specific social groups or classes that could serve as the agents of historical change, as the working class had once done within socialism. Her essay's final, gun-slinging line – 'I'd rather be a cyborg than a goddess' – said it all.

Though the term does not appear in the essay, *A Cyborg Manifesto* is often cited as an example of post-humanist philosophy, which is an outgrowth of the more general discipline, or tendency, known as postmodernism. Certainly Haraway's determination to debunk the dualisms and categories of Western post-Enlightenment thought is very close in style and substance to the French philosopher Michel Foucault. In his 1966 book *The Order of Things*, Foucault argued that the concept of 'humanity' was a creation of the late eighteenth century – a social and ideological construction born of new relations of power and ways of thinking about the world. Like Haraway, he regarded *Homo sapiens* as a fiction – 'an invention of recent date', and quite possibly one 'nearing its end'.[18] In a sense, *A Cyborg Manifesto* presents the fulfilment of this quasi-prophecy: Haraway is describing *and celebrating* the fact that 'the human' has ceased to exist as a category.

When challenged on these arguments, Haraway would often double down on the claim that her cyborg was merely a metaphor, though it is difficult to see what force, if any, the metaphor would have if it didn't correspond to a particular claim about reality, the claim in this case being that human beings are in some material sense *post*-human. In any case, many of the academics and activists who have taken up the cyborg banner are not thinking metaphorically at all. This is especially obvious in contemporary gender studies. In *Full Surrogacy Now* (2019), critical theorist Sophie Lewis argues for a future utopia based on 'gestational communism': a society in which everyone can become pregnant, and 'gestational labour' and child-rearing are shared. Drawing on Haraway's essay, Lewis is more than comfortable with using transformative technologies to redistribute such labour, and Haraway's cover blurb – 'The seriously radical cry for full gestational justice that I long for' – suggests that she's comfortable with it too.

Though rarely as radical as Lewis's prescriptions, the idea that sex and gender are fictions – indeed, that the *human being* is a fiction, constructed through 'discourse' and ideology – has found its way from academia and into the wider culture in recent times, albeit in an often diluted form. One significant step along this path was the publication of Judith Butler's 1990 book *Gender Trouble*, which suggested not only that gender was 'performed' according to conventional 'scripts', but also that our notion of biological sex as in some sense 'prior to' such gender performances – as something that can be separated from gender – was itself a cultural construct. While conservatives might once have said that one's biological sex defined one's gender (which it doesn't), Butler was asserting that conventional notions of gender *actually constructed sex* – that there was no category of 'biological sex' from which ideology and discourse were absent. As she put it:

> Can we refer to a 'given' sex or a 'given' gender without first inquiring into how sex and/or gender is given, through what means? And what is 'sex' anyway? Is it natural, anatomical, chromosomal, or hormonal, and how is a feminist critic to assess the scientific discourses which purport to establish such facts for us? [...] If the immutable character of sex is contested, perhaps this construct called 'sex' is as culturally constructed as gender; indeed, perhaps it was always already gender, with the consequence that the distinction between sex and gender turns out to be no distinction at all.[19]

Though Butler is often misrepresented as saying that biological sex doesn't exist, her argument that it is impossible to separate the concept of sex from its cultural construction has buried itself deep in the minds of many activists, and in the minds of many who support them in their struggle. On the young left in particular the notion that there is no implicit relationship between one's gender and one's biological sex has become an article of faith. Indeed, and more generally, any mention of biology, at least as it relates to human nature, is likely to be regarded as 'essentialist' – as reductive in the Social Darwinist sense. Sex and sexuality have catalysed a more general suspicion of the body as something through which we come to know the world.

No doubt the motives are admirable: sexuality, after all, is one of the main channels through which right-wing reactionaries press their case, with appalling consequences for those who fall outside the sanctioned categories. The problem, however, is that this political stand is inadvertently breathing life into a view of human beings as in some sense *above* or *apart from* nature. And it is here that post-human philosophy may one day find – *may already be finding* – common cause with the transhumanism of Kurzweil et al. For while post-humanism and transhumanism are ostensibly very different phenomena – the first a philosophical rejection of post-Enlightenment humanism, the second an enterprise aimed at bringing humanism into the cyber age – it is plain that both intellectual tendencies see the body as highly malleable, if not exactly expendable. They are both invested in Project Cyborg.

Project Human

In her 2019 novel *Frankissstein*, the British author Jeanette Winterson moves between the different worlds of info-tech and gender politics – of transhumanism and transexuality – in a meditation on the values of both milieux. As its title suggests, the novel is in dialogue with Mary Shelley's *Frankenstein*, and Winterson's narrative is split between an account of that great book's composition on the shores of Lake Geneva in 1816 and our own technoscientific moment. The characters in the contemporary setting ghost the characters in the older one: Victor Stein (Victor Frankenstein) is a celebrated professor working on AI and cryogenics. Ry Shelley (Mary Shelley) is a transgender doctor. And Ron Lord (Lord Byron) is the oafish purveyor of anthropomorphic sex robots.

The novel is a lot of fun, but its underlying thesis is an extraordinary reversal of Shelley's anti-Promethean thrust. For though the author is rightly concerned about sex dolls that reproduce misogynistic fantasies of female sexual availability, it is clear that she also regards the use of transformative technologies to reengineer one's body in accordance with one's 'real' self as a (political) cause for celebration. As Ry puts it: 'I'm a woman. And I'm a man. That's how it is for me. I am in a body that I prefer. But the past, my past, is not subject to surgery. I didn't do it to distance myself from myself. I did it to get nearer to myself.'

This is perhaps a rather simplistic take on what a transgender person might go through, emotionally and philosophically, before deciding to take such a momentous step. But as an expression of a certain contemporary belief in the link between bodily transformation and personal freedom, it is undeniably of our time. In assertions such as Ry's we glimpse the outline of an entirely new relationship between humanity and technology, based on a view of the human body as some sort of inefficient encumbrance, or at least something less real than one's 'true' identity. Leaving the specific issue of gender-reassignment (or confirmation) surgery to one side, it is clear that Winterson is channelling a very modern idea of liberation. For those who might share that idea of liberation, I want to pose the following question: is the 'self' we think we are travelling towards really so easy to identify?

We live in a world in which technoscientific capitalism is making identities fragile and fluid. As opportunities for embodied sociality decline, we are told to 'get out there' and distinguish ourselves – 'to be true to ourselves', whatever that means. At the same time, and in this spirit of distinction, the body is increasingly remade through surgeries such as breast augmentation and buttock enhancement – surgeries that, in the case of a Kim Kardashian (half human being, half bouncy castle), are no less 'pornographic' than the sex dolls peddled by Ron Lord in *Frankissstein*. Body modifications and biohacking are likewise rising rapidly. Are such phenomena free from ideology? Is our regard for individual choice now such that we celebrate cosmetic self-mutilation in the search for our authentic selves? And might there be some relationship – however complex, however deep – between these cultural phenomena and the politics of a new generation of progressives?

'Biology is not destiny,' writes the British feminist Helen Hester, in her book on Xenofeminism – a movement that promotes the use of technology in order to abolish gender. She's right: the human animal is far more interesting than the biological reductionists would suggest. But culture is not exclusive destiny either, and if we begin to behave as if it is, treating our bodies as so much stuff to be manipulated, we will surely be setting ourselves up for trouble. The body is not an extension of the self; it *is* the self, and the self is *it*. To consistently treat it as something expendable, or infinitely malleable, is an enterprise no less dangerous in the hands of the Promethean left than it is in the hands of the info-tech wizards. It is expressive of the same disorder – a secular faith in progress. It is a human catastrophe in the making.

7

SOME SWEET OBLIVIOUS ANTIDOTE

Pharmaceuticals and Human Flourishing

In some ways it's surprising that the founder of transhumanism and the author of *Brave New World* were brothers. Born in 1887 and 1894 respectively, Julian and Aldous Huxley were both enthusiastic about science, up to and including the science of eugenics. Grandsons of the great Thomas Henry Huxley, also known as 'Darwin's Bulldog', they were liberal intellectuals of a recognisable sort: polymathic, broadminded, rational to a fault. And yet while one of them kick-started an intellectual movement dedicated to the proposition that humankind must 'transcend itself', the other wrote a novel depicting the horrors of technological progress. What explains this apparent discrepancy?

I may be overreaching, but I like to think the answer relates to the fact that there is something about literature that rebels against the idea of the perfect society, and of human perfectibility in particular. Perhaps with only one exception – William Morris' *News from Nowhere* (1890) – utopias do not work as fiction, for the simple reason that a perfect society, being a human impossibility, is necessarily devoid of human interest. Even the mighty H. G. Wells – a utopian rationalist if ever there was one – is better known for his mad vivisectionists and tentacular Martian 'fighting-machines' than he is for *A Modern Utopia* (1905), a thinly disguised philosophical essay with all the narrative drive of a takeaway menu. As Huxley intuited, there is something *implicitly* self-defeating about the idea of a perfect society. The word 'dystopia' is often taken as an antonym of 'utopia', but in the pages of *Brave New World* we are reminded that the distance between the two jurisdictions is not as great as we might imagine, and might even be no distance at all. If *Brave New World* continues to resonate in a way that *Nineteen Eighty-Four* (1949) does not, this is part of the reason: it reminds us that *all* utopias are pregnant with their own negation – that all utopias are dystopias in vitro.

The great twentieth-century dystopias were not intended to be prophetic texts. They are satires, or 'social formula' novels, creatively removed in time or space, no more in the business of making predictions than John the Baptist or the Oracle of Delphi were in the business of writing literary fiction. But they continue to resonate nevertheless, and often in interesting and fruitful ways. Take, for example, the attitude to technology evinced in Huxley's *Brave New World* as compared to *Nineteen Eighty-Four*. In Orwell's totalitarian vision, technology is merely a means to an end, the end being total political power. Hence the bidirectional 'telescreens' – tools of both propaganda and surveillance – installed in every citizen's apartment. But in *Brave New World*, technology is used not merely to police the populace but to mould it into compatibility with the World State's guiding principles of 'community, identity and stability'. In the interests of maximum efficiency, genetic-engineering technologies are employed to manufacture human beings and to sort them into different classes; but the system is as dedicated to emotional wellbeing as it is to rational organisation. As the Director of the Central London Hatchery and Conditioning Centre explains to a colleague, 'All conditioning aims at ... making people like their unescapable social destiny.' This is what makes the World State so sinister: its leaders, the Controllers, seem to genuinely care that its citizens are not just productive but *happy*.

More specifically, its sinister character derives from the fact that citizens are *engineered* to be happy, and this goes to one of the most interesting features of Huxley's futuristic vision: the use of social conditioning and pharmacological technologies to stave off feelings of discontent and dejection. All dystopias feature some kind of 'antidote' that is used to reconcile individuals to an inhuman system. The music and pointless chatter of mass media in Ray Bradbury's *Fahrenheit 451* (1953) operates as such an antidote, as do the televised gladiatorial spectacles in Suzanne Collins' *Hunger Games* trilogy (2008–2010). So too do the Victory Gin and pornography distributed among the masses in *Nineteen Eighty-Four*, and the addictive 'wireheading' technologies in various cyberpunk scenarios. In Huxley's vision, by contrast, the favoured antidotes are not technologies of checking out but technologies of fitting in. Anyone experiencing feelings of unhappiness is encouraged to take the drug soma, the benefits of which are set out by the Resident World Controller of Western Europe, Mustapha Mond: 'And there's always soma to calm your anger, to reconcile you to your enemies, to make you patient and long-suffering. In the past you could only accomplish these things by making a great effort and after years of hard moral training. Now, you swallow two or three half-gramme tablets, and there you are.'[1]

This is something quite different to the escapist antidotes mentioned above. Soma is not a sedative, and certainly not a psychedelic; it is a pharmacological preparation aimed at social integration. It is like the Penfield Mood Organ in Philip K. Dick's dystopian novel *Do Androids Dream of Electric Sheep?* (1968), a bedside device that allows individuals to simply dial up an alternative mood, and the various drugs that appear on screens in the 2006 sci-fi thriller *Children of Men* (including Bliss, for anxiety and depression). It is a tool of *enhancement* that brings individuals into line with the sanctioned attitudes and behaviours of the system.

In this sense, it is clear that *Brave New World* does indeed have something important to say about our own society. For while escapist drugs like alcohol have been a feature of human communities for millennia (by some reckonings, biotechnology itself began with the fermentation of beer: a breakthrough, in the opinion of this writer), the emergence of such 'pragmatic' antidotes is a relatively recent phenomenon. The use of soma in *Brave New World* has been likened to the craze for the tranquiliser valium in the 1960s and 1970s.[2] But in fact it appears to share the properties of modern antidepressants as well, and so raises questions about whether those drugs act as 'techno-fixes', circumventing the need for a public debate about why depression appears to be rising in certain societies. In the market race to create antidepressants, as well as stimulants and amphetamines for children with conditions such as ADHD, these questions have often been ignored, not least because the driving ethos of technoscientific capitalism decides the issue in advance by treating society as a collection of autonomous, machine-like individuals in need of ongoing maintenance. It is therefore important to ask what function such drugs perform, and whether they might be encouraging us to effectively reengineer ourselves in line with a system that, though no World State, is in certain fundamental ways unconducive to our deepest needs.

In this chapter I'll dig down into those questions, and consider whether we are witnessing a reversal of the direction of fit between human beings and the societies they've created, or that have been created on their behalf – that instead of creating societies that engender happier human beings, we are in the process of creating human beings that fit into the society we happen to have made. If that is even partly true in the case of mood-altering medications, then we are indeed beginning to turn ourselves into something like (chemical) cyborgs, and not in the spirit of 'transcendence' dreamed of by Julian Huxley and his transhumanist descendants, but in line with the values of the very system that made us feel like shit in the first place – with the values of technoscientific capitalism. A brave new world indeed.

The Economy of Melancholy

In the final act of Shakespeare's *Macbeth* (1623), the Scottish nobleman implores his physician to rid his wife of her mental distress by way of 'some sweet oblivious antidote'. The doctor replies that he cannot do so, and that in cases such as Lady Macbeth's 'the patient / Must minister to himself'. His response is of course the proper one. Lady Macbeth's emotional turmoil is a reaction to her circumstances, which in this case happen to be of her own making. She is 'Not so sick,' as the doctor puts it, 'As she is troubled with thick coming fancies, / That keep her from rest.'

In modern psychiatric terms, Lady Macbeth's emotional condition is 'reactive': it is not biological in origin but a response to a stressful or traumatic event. Nevertheless, it is highly likely that were she to present to a modern GP, she would be offered 'some sweet oblivious antidote' in the form of a course of antidepressants. She might be offered counselling, too, or even a referral to a psychiatrist; but her GP would almost certainly *raise* the topic of psychotropic medication. Nor would the doctor be wrong to do so. By definition a generalist who relies on researchers and medical specialists to decide what medications are appropriate, a GP must be guided by the consensus in those fields, as well as by the responsibility to reduce the suffering of her patient if possible. Since such drugs have been shown to lift low mood, significantly reduce anxiety and even put paid to those 'thick coming fancies' that today we label 'negative thinking' (or so the companies that make them claim, not always uncontroversially), a pill would most likely be in the offing.

Today there are countless such pills, and for anyone over a certain age their multiplication has been a staggering phenomenon. In the space of just fifty to sixty years, antidepressants have become ubiquitous – an inescapable modern theme. The arrival of Prozac in the late 1980s and 1990s, which largely retired the so-called tricyclics that had reigned supreme in the previous two decades, was especially significant, being the point at which such medications decisively entered the wider culture, as something between a medication and a lifestyle drug. As the first of the SSRIs, or selective serotonin reuptake inhibitors, Prozac also played a role in convincing both medical professionals and the general public that the big pharmacological companies were closing in on the mystery of depression. Describing both the cause of depression (insufficient serotonin) and the method by which it would be rectified, it not only gave users confidence that the pharma companies knew what they were doing, but also served to reassure them that depression was an organic condition, unrelated to all those personal 'failings' on which they were apt to dwell. In fact, there are many remaining questions about precisely how antidepressants work, but the view that depression has a neurochemical cause and requires a neurochemical solution is now hegemonic. Accordingly, prescriptions for antidepressants – for both SSRIs and SNRIs, which also target norepinephrine – have risen, and are rising, steeply. In Australia, for example, more than 4.7 million prescriptions for Zoloft (the world's most popular SSRI) were issued between July 2019 and June 2020.[3] As of this writing, one in seven Australians is taking antidepressant medication.[4]

Taken by itself, such a rise in prescriptions is not necessarily a worrying development. No doubt the pharma companies would say that their products are a remedy for conditions that, in the past, were under-diagnosed or treated ineffectually, so prescriptions will increase in line with our expanding understanding of those conditions. This is certainly a possibility. Modern medicine has brought enormous relief to sufferers of all kinds of conditions, including conditions of the mind, such as bipolar disorder and schizophrenia. From the bloodletting practised in medieval times to the hellish asylums of the early modern period, sufferers of severe depression have often been on the receiving end of barbaric medical practices; but since the middle of the twentieth century, great strides have been taken in the management (and it is only ever the *management*) of clinical or 'endogenous' depression. As someone who grew up with a clinically depressed mother who was hospitalised at several points in her life and subjected to many different treatments, I have no doubt at all about the necessity of medication. In fact, I'm convinced that if it hadn't been for drugs my mother would not be alive today.

The problem is that, when considering milder forms of depression, it is far from clear those more common ailments are organic or biological in the way the pharma companies suggest. Depression rose sharply during the global debt crisis, and again during the COVID-19 pandemic, suggesting that the increased prescription of antidepressants in those periods was answering to emotional crises that were to some extent social in origin.[5] This is not to say that GPs were wrong to prescribe those antidepressants. It is just to say that in making sense of what appears to be a more *general* increase in the incidence of depression and anxiety, we must allow for the possibility that certain aspects of our society are increasingly 'depressogenic' – that they *cause* depression and anxiety.[6] Add in the fact that depression seems to be rising faster in developed countries than it is in developing ones, and we are confronted with the possibility that there is something in the character of liberal-capitalist societies that is unconducive to human happiness, or to what we think human happiness should entail.[7] That last qualification is an important one, for it could be that it is our expectations of what a fruitful modern life should offer, or even our notions of what depression *is*, that are changing, rather than the incidence of depression itself. And of course it could also be the case that the discrepancy between Western and non-Western countries is partly cultural in origin, with individuals in more 'traditional' societies less likely to admit to feelings of despair. The caveats are manifold. But even accepting all of them, we are still left with societies in which the *claim* to be unhappy is increasing, and in which that claim is increasingly answered with drugs.

Personally, I don't think there's much doubt that these developments are related to a real rise in mental distress. The COVID-19 pandemic was a reminder to all of us of how fundamental to human flourishing the embodied presence of others is, and there's no question that the spike in anxiety and depression that attended the lockdowns was related to that need. Why, then, would we expect a society that has turned housing into a financial instrument (undermining community life in the process), that incorporates screen technologies into every level of social life, and that esteems the values of competitiveness and striving above cooperative endeavour and social solidarity *not* to engender an increase in feelings of meaninglessness and despair? Could it be that some instances of what we label 'depression' are in fact a social-psychological disorder related to the steady erosion of social solidarity in hi-tech neoliberal societies? And if so, how do we feel about using psychotropic medication to bring those who are struggling up to speed?

There is a long tradition within social science of thinking about the relationship between modernity and human flourishing. Modern sociology itself can be said to start with Émile Durkheim's comparative study of suicide in traditional and non-traditional communities. Published in 1897, *Suicide: A Study in Sociology* grew out of Durkheim's longstanding interest in the struggle of modern societies to maintain social integration and solidarity in the face of fewer traditional bonds and the rise of new institutional forms. Its principal (and controversial) finding – that suicide was less prevalent in Catholic communities than in Protestant communities – appeared to suggest that higher levels of social integration among Catholics accounted for this discrepancy. Durkheim's nephew, Marcel Mauss, attacked the question from another angle, but lent credence to his uncle's conclusion that modern industrialised societies were increasingly prone to 'anomie' – the fraying of the social fabric resulting from a breakdown in conventional standards, codes, morals and behavioural norms. In his seminal 1925 essay *The Gift*, Mauss dug down into the anthropology of gift-giving practices in 'primitive' societies, unearthing a richly symbolic field in which acts of exchange and reciprocation functioned to bind communities together in ways that commodity economies, by their very nature, could never replicate. As Karl Polanyi would argue in his 1944 book *The Great Transformation*, it is in the nature of liberal-capitalist societies to lift economic relationships of exchange out of their previous social contexts and to transform human thinking in the process, idealising transactions as 'rational' rather than reciprocal.[8] Over time this means that traditional practices are supplanted by formal institutions that promote a market economy, and indeed that 'the economy' becomes in some way separate from other aspects of social life – a sphere in which people 'go to work' and interact with others in distinctive ways. The result, again, is an overall decline in social solidarity, as relationships become increasingly abstract and life-worlds more precarious.

Because much modern commentary is characterised by a certain neophilia – a reflexive regard for the new, combined with a tendency to dismiss any opinion that some things may be getting worse as Golden Age thinking or simple nostalgia – these traditions of intellectual critique are not as prominent as they were in the past. And yet, for all the gains that have been made, and are yet to be made, in the name of modernity – education, mobility, travel, diversity – it is clear that social solidarity, and the emotional benefits that flow from it, are in decline. Even a few decades ago most people lived in much closer networks of kin and community, grouped around extended family, neighbourhood, congregation and so on. Such a world could be stifling and prejudiced, but it was nevertheless the guarantor of a certain sense of (stable) selfhood, deriving from one's position in the family, one's job, one's parish, one's union, one's pub, one's local team and one's sense of place. To some extent this world was held in place by the thirty years of social democracy that followed the end of World War II, especially in European countries, which, with the Soviet Union to their east, knew that they could no longer take the sacrifices of working-class communities for granted. But in the past few decades of neoliberalism, those working-class networks have been swept away as comprehensively as the physical neighbourhoods they dwelt in, while the ethos of meritocracy – a word coined by the sociologist Michael Young, in the spirit of satire, not recommendation – means that those who do less well economically are left (in Young's phrase) 'morally naked', no longer able to attribute their 'failings' to ill-luck or lack of opportunity. Even in the wake of the global debt crisis – a crisis caused by financial capitalism – this lack of a social lens was evident. As the Marxist geographer David Harvey observed in *The Anti-Capitalist Chronicles* (2020), foreclosed-upon Americans, when asked about their predicament, were more likely than not to blame themselves.[9]

At the same time as the material conditions for social solidarity have frayed in liberal-capitalist societies, a highly individualised idea of what a 'happy life' might entail has also become conspicuous. This idea owes as much to the counterculture of the 1960s and 1970s as it does to the architects of neoliberalism. But over time the more pastoral and introspective elements of that great social and generational upheaval have fused with the priorities of neoliberal capitalism. The self-help boom of the 1980s was one aspect of this synthesis, in which the idea that one had a *right* to be happy was transformed into something more like a *duty*. To some extent, this new dispensation was the invention of an increasingly sophisticated advertising industry. As Christopher Lasch noted in *The Culture of Narcissism* (1979), the main 'product' sold by this industry was not a new car or a brand of cigarettes but an idealised version of the consumers themselves.[10] Having made human identity more precarious through the steady erosion of social life, capital thus proposes an antidote in the form of a sexier, more self-confident one.

Combined with those technologies of absence discussed in Part I – technologies that engender a 'morbid' sociality and ever more fragile identities – these factors have led to a society that can sometimes seem *designed* to cause as much distress as possible without subjecting its members to actual violence. Whether we choose to call that distress 'depression' is neither here nor there. As only conservative commentators of the stiff-upper-lip school seek to deny, the distress is real, and young people in particular are feeling increasingly anxious and dejected. A temporary inability to cope with the pressures and demands of contemporary life is now a ubiquitous phenomenon.

What should we do about it? A sensible response might be to look at the society we've created and think about what we need to change: to think about how we design communities, the value we place on work and creativity, the pressures on the modern family, the instrumental character of modern education, and our increasing alienation from nature.

Or we could treat the problem in a way that reproduces the values of individualism and technophilia that led to it in the first place.

No prizes for guessing which path we've chosen.

The Great Reversal

'The junk merchant doesn't sell his product to the consumer,' wrote William S. Burroughs in *Naked Lunch* (1959), 'he sells the consumer to his product.' Swap out 'junk' for 'antidepressant' and this statement still holds substantially true.

The claim is not as militant as it seems. For in one sense all capitalism works in this way. As Vance Packard showed in *The Hidden Persuaders* (1957), advertising and marketing companies are not the servants of demand but the masters of it, creating desire among consumers for products they didn't know they wanted. More recently, and as we saw in Chapter 1, the 'junk merchant' aspects of capitalism have moved into a different phase, as advertising companies have adopted a model *explicitly* designed to 'hack' the brain. In the era of info-capitalism especially, Burroughs' comment has a general relevance.

There is, though, something particularly egregious about this process as applied to a medicinal product, especially one aimed at depression and anxiety. If I'm right that capitalism is partly responsible for the sense of emotional drift and dejection that characterises modern life in developed societies, then the spectacle of massive corporations competing to sell us back the peace of mind we have lost does begin to look a little, well, sick, especially when you factor in the well-documented instances of sharp business practice that have accompanied that race for market share. The junk merchant, writes Burroughs, does not improve his merchandise; he 'degrades and simplifies' his client. To the extent that it peddles antidotes for anomie and alienation – for the 'degradations' of modernity and neoliberalism – the trade in psychotropic medication does have a hint of the crack-house about it.

The rise of antidepressant medication is related to an important shift in emphasis within the psychiatric profession. In the 1960s and 1970s, radical psychiatrists like R. D. Laing had tended to treat mental illness and psychosis as primarily social phenomena, sometimes making wild claims to the effect that conditions such as schizophrenia were 'crazy' responses to a crazy world. Miloš Forman's hugely popular movie *One Flew Over the Cuckoo's Nest* (1975), based on Ken Kesey's novel of the same name, set out the popular version of this view, which was itself a kind of antidote to the chemical and electroshock methods favoured in psychiatry at that time. Under fire from this radical challenge, mainstream psychiatry became even more attached to a biological vision of depression, finessing its diagnostic categories in a way that led to a more descriptive approach to mental illness overall – an approach that tended to exclude the critical and interactive aspects of treatment. Increasingly, mental pathologies were constructed around collections of symptoms, which could be shown to respond to different drugs. Mental illness was becoming a condition of the *brain*, rather than a condition of the *self*.

As the psychotherapist Gary Greenberg demonstrates so brilliantly in *The Book of Woe* (2013), such 'descriptive psychiatry' goes hand in hand with the phenomenon of 'diagnosis inflation'. Investigating the evolution of the *Diagnostic and Statistical Manual of Mental Disorders* – the American Psychiatric Association's taxonomy of psychological ailments, first published in 1952 and currently in its fifth edition – Greenberg shows how intellectual hubris on the part of the psychiatric profession and the market in prescription drugs combined to swell the DSM-5. Categories such as 'prolonged grief disorder' (which can be diagnosed from six months after a personal loss) and 'disruptive mood dysregulation [*sic*] disorder' have proven especially controversial, referring as they do to emotional states (bereavement) and patterns of behaviour (angry outbursts) that most of us understand intuitively as natural aspects of ordinary life. As the psychoanalyst Darian Leader has argued, one reason for this widespread pathologisation is that new rules in the 1960s obliged manufacturers of novel drugs not only to list active ingredients, effect periods and so on, but also the specific diseases and conditions they proposed to treat. This led drug companies to market the *disorder* alongside the medication, as when the NFL running back Ricky Williams was recruited by a PR company to talk about his 'social phobia' (shyness) on *The Oprah Winfrey Show*, prior to spruiking GlaxoSmithKlein's new SSRI, Paxil CR. At the same time, it led them to create new categories of anxiety and depression in line with new drugs, which may have only subtly differing profiles. The human reintegration that psychiatry is traditionally supposed to promote is replaced by an ever-expanding taxonomy of disjunctive symptoms and 'abnormal' behaviours.[11]

The effect of this is to confound the difference between normal and abnormal mood states. Mild depression and anxiety come to be seen not as reactions to events in the world, but as disorders – as medical conditions – in a way that forecloses any thoroughgoing discussion of depression as a social phenomenon related to changes in society at large. Where lip service is paid to social phenomena such as social isolation or meaningless work, they are subsumed within a broader public-health model that is both highly medicalised and highly individualised. Meanwhile, individual sufferers, who may be relieved to hear that their anguish has a neurochemical origin – that their feelings of failure are only *feelings* of failure – are at once denied any genuine insight into what may be an emotional response to conditions in the 'real world' and invited to regard themselves as systems rather than selves or souls – as aggregations of the various chemicals coursing through their physical bodies. What might in other circumstances have led to a more textured analysis of one's life – and, at the level of social policy, to a conversation about how that life is in some ways fundamentally ill-served by the trajectories of contemporary society – is reduced to a discussion about neurotransmitters.

In treating the issue of mental illness in this way, modern psychiatry reproduces the individualised and informational outlook of technoscientific capitalism. As a system in which science is subordinated to technology and technology is subordinated to the profit motive, technoscientific capitalism is apt to reduce the individual to an autonomous biological system, as opposed to a socially constituted self. So the pharmacological approach to conditions such as anxiety and depression is pregnant with the very values that may have caused, or partly caused, the anxiety and depression in the first place.[12] Even non-pharmacological approaches, such as cognitive behavioural therapy and rational-emotive therapy, will often channel this view of the world, encouraging clients to re-organise their thought 'patterns' as if they were so many parts of a program that had temporarily ceased to work as it should.[13] The self-help books that invite their readers to 'rewire' or 'reprogram' their brains for success are at the trashy end of this phenomenon, but are based on similar assumptions – assumptions that grow from a view of human beings as machine-like entities in need of the occasional tune-up. At an even more mundane level, 'life skills' such as personal resilience, mindfulness and even empathy are now abstracted from their social contexts and introduced to schoolchildren or employees as psychological competencies to be mastered. Such 'social technologies' speak to the *lack* of such qualities in hi-tech modern societies; and yet we continue to address that lack at the level of the individual rather than society.

The *reductio ad absurdum* of this situation, though not perhaps its most sinister iteration, is the recent appearance of mental health apps such as MoodMission, BetterHelp, MoodFit, Bearable, Shine and Happify, all of which help you keep track of your moods and provide tools for addressing 'negative' emotions. Most of these apps are algorithmically based, which doesn't inspire confidence, though Alison Darcy, the CEO of Woebot, regards this as a selling point. 'Robots are not going to judge you,' she explains. 'We've removed the stigma by completely removing the human.'[14] The statement will strike many as self-satirising, a contradiction of one of therapy's key purposes – to reintegrate individuals back into the society from which they feel alienated. As the philosopher Mark Furlong has suggested, such chatbots are as likely to summon 'self-concern, vulnerability, entitlement and the valorisation of convenience' as they are camaraderie and mutuality.[15] Darcy's confidence that the absence of humans is good for the user because it eliminates 'stigma' suggests that the act of seeking help from another human being is emotionally fraught. But of course it is in reaching out to others that we take the first tremulous steps towards healing. The idea that a bot – on a *smartphone*, of all things – could stand in for that process is absurd and dangerous.

It is dangerous because despair and depression, as anyone who has experienced these states will know, are characterised by the flight of meaning from the world; work, friends and even family cease to bear in on us as they should, as sources of emotional enrichment, as the world becomes, in Hamlet's words, 'weary, stale, flat, and unprofitable'. To respond to such a flight of meaning by inviting people to regard themselves as 'automata that can regulate their states by modifying their inputs' (Furlong) may have short-term positive consequences; but it is also likely to reproduce that lack of meaning over the long term, while also ignoring its deeper causes.[16] The same goes for psychotropic drugs. To say that a chemical treatment is effective is not to say that the condition it is treating is necessarily chemical in origin. That would be to mistake a stage in the process of becoming unwell for its starting point, its original cause. I have a scar in the middle of my forehead, but it wasn't caused by a tear in my epidermis. It was caused by an act of violence. The point I've been making about modern depression is that it is caused by a kind of violence, too, and that simply stitching ourselves back together is no substitute for safer streets.

No, we do not live in a dystopia, and antidepressants are not soma – not yet. But we should allow for the possibility that in using pharmacological technologies to help us cope with modern society, we are preparing for a future in which our deepest needs – for reciprocity and conviviality, creativity and connection with nature – are likely to remain unmet, and that this may be the beginning, not the end, of a process that goes on for a very long time. At the very least, it is worth beginning to think about the effect that process is having and where it might be leading us.

Moral Enhancement

Just a few months into the COVID-19 pandemic, as jurisdictions across the world were grappling with the challenges presented by the novel coronavirus, an article appeared on *The Conversation* suggesting a (novel) line of defence. Written by Professor Parker Crutchfield, an ethicist at Western Michigan University, it suggested that part of the response to future pandemics should be to develop new medications that would instil in people a greater sense of social solidarity, so that they would be more inclined to comply with mask-wearing and lockdown mandates. Invoking the 'tragedy of the commons', which describes the human tendency to act in our own self-interest and flout rules designed to benefit the majority, Crutchfield lamented the recalcitrance of certain elements in the United States and elsewhere, whose selfishness was frustrating public-health policies aimed at containing COVID-19. The solution, he suggested, was 'moral enhancement'. '[L]ike receiving a vaccine to beef up your immune system,' he wrote, in a questionable analogy, 'people could take a substance to boost their cooperative, pro-social behaviour.'[17] Lest he be dismissed as a fantasist, Crutchfield was careful to concede in advance how challenging such a proposal might sound. The plan, he averred, would be 'controversial'.

'No shit!' would be an economical way to summarise the comments that unspooled beneath the article. This was published on a website designed to give academics a platform to speak to non-academics, which in the last decade has gained a huge following from knowledge-class readers, eager for opinion articles that can act as a levee against the ever-rising tide of 'fake news' that characterises the digital era. That such an article could appear in such a forum was disconcerting, to say the least, and some respondents were clearly angry about the editorial decision to let it appear – less because of its moral illiteracy than its scientific and political implausibility. Not since Jonathan Swift suggested that the Irish poor fricassee their own sons and daughters had such a barmy suggestion been made.

Nevertheless, Crutchfield is not alone in his ambition to reengineer human morality. 'Moral bioenhancement' (MBE) has been a hot topic in bioethics since the Oxford philosopher Thomas Douglas introduced the idea in 2008, in an article in the *Journal of Applied Philosophy*, and its advocates have been unapologetic in making the case for such intervention.[18] In *Unfit for the Future* (2012), Julian Savulescu and Ingmar Persson argue that human morality, having evolved amid small communities with rudimentary technologies, is unsuited to modern societies, and we should be thinking about ways to overcome this limitation using MBE. Subtitled 'The Need for Moral Enhancement', their book is short on scientific detail but convinced of the case for developing technologies – most probably pharmacological technologies or biotechnical interventions – that compel human beings to behave as a 'garden-variety virtuous person' would behave. As the reactions to Crutchfield's article would suggest, the advocates for MBE still have all their work ahead of them in terms of convincing the general public that this is something worth considering. But the mostly utilitarian philosophers who have declared themselves in favour of it are far from marginal figures, and their outlook does appear to chime with the broader transhumanist view of the world evinced by Ray Kurzweil and his equivalents.[19] There is nothing to say that we won't come round eventually to their way of thinking.

There are both philosophical and practical objections to the idea of moral bioenhancement. The philosophical objections are clear enough, and can either be framed as a series of questions – Who is this garden-variety good guy? How is a decision moral at all if it isn't arrived at through the exercise of reason? Who the hell do you think you are? – or, more positively, as an assertion that morality cannot be neatly extracted from the social, cultural and historical processes through which it is formed, challenged and re-formed. Moral reasoning is not some cool bit of software that is uploaded to the brain and allowed to run; it is an ongoing, peculiarly human activity performed in negotiation with others. As both Aristotle and Immanuel Kant maintained, it is the capacity that differentiates human beings from plants and nonhuman animals, inextricably tied to the freedom we have, and to the dignity we accord ourselves. To reduce it to whatever chemical trace it deposits in the folds of the brain is reductionism triple-distilled.

The practical objections to the proposal are perhaps easier to overcome. In fact, the more one looks into the question, the more credible MBE comes to seem, notwithstanding that any 'bioenhancement' that did occur would not be moral but simply biochemical, for the reasons stated in the previous paragraph. For example, it appears that some antidepressants and anti-anxiety medications affect a number of the specific brain chemicals involved in moral decision-making.[20] According to neuroscientist Molly Crocket, who has conducted her own research on the topic, there is even evidence that the antidepressant citalopram and the anti-anxiety drug lorazepam affect responses to the 'Fat Man' iteration of the Trolley Problem discussed in Chapter 3. After taking citalopram, respondents appeared to be *more* likely than the placebo group to push the fat man off the bridge, while those who had taken lorazepam were *less* likely than the placebo group to do so.[21] As Crocket notes, the question of whether either medication could form the basis for a 'morality pill' depends entirely on what your moral position is: trolley problems have no right answers, after all. But in the event that we all decide to agree on what is (broadly speaking) morally right – or, more likely, that someone decides what is morally right on our behalf – the potential for some modicum of 'moral engineering' appears to exist in the drugs we have already. In the event that a pharmaceutical company was permitted to pursue such a product line *regardless of whether we agree or not* – an even more likely scenario – I imagine the question of whether to pop the morality pill or remain unenhanced would be framed in terms of consumer choice: technoscientific capitalism's ace in the hole.

In that case, the emergence of such pharmaceuticals would be an uneven and chaotic process, subject to squabbles over patents and approvals, but trending towards the same transformation of social life that is currently occurring with respect to antidepressants. The MBE utilitarians would never get their stated wish for a widespread public conversation about what is and what isn't moral, but their reductionism would work to swing the public mood in favour of the companies manufacturing the new pharmaceuticals. Slowly but surely, we'd grow used to the idea that morality, like happiness, can be engineered.

And at that point? Well, we would have succeeded in turning Huxley's nightmare vision into a reality. Armed with sweet oblivious antidotes not only to our anxieties but also to our moral failings, we would have turned ourselves into chemical cyborgs – not Kurzweil's disembodied consciousnesses but embodied techno-organisms.

Psychotropic zombies, if you will.

Monsters, more or less.

CODA

A Brain the Size of a Planet

In December 2022, as Elon Musk was publicly bickering with users on his newly acquired platform, a technology story of much greater consequence spread quickly through the mainstream media. The research 'laboratory' OpenAI had released its new chatbot, ChatGPT, and it was ... well, it was remarkable. A vast, multi-levelled learning program, ranging in recursive fashion across a gigantic database in order to retrieve and synthesise information and reproduce natural language structures, it could interact with users with the uncanny humanness of a HAL 9000 or a C-3PO. Need an essay on Rachel Carson's *Silent Spring* and its influence on the New Left movements of the 1960s and 1970s? Or a report on the 1978 World Cup final as related by Marvin the paranoid android in *The Hitchhiker's Guide to the Galaxy*? What about a poem in the late style of T. S. Eliot, on the subject of an ageing Lothario who can't stop thinking about making love to his neighbour? ChatGPT would be happy to oblige. Proving the point (and proving as well that perhaps they weren't as essential as they'd imagined), an army of journalists flew into print with articles wholly or partly composed by the new artificial intelligence. For the most part, one couldn't spot the difference.

As ChatGPT became so overloaded with requests from users that its servers crashed, the question of whether its striking affordances amounted to intelligence proper bubbled away in the media commentary. For those still devoted to the Turing Test, it clearly did. A decade before, in 2012, the scientist and AI researcher Ben Goertzel had proposed a new version of the Turing Test, stating that an AI capable of obtaining a degree in the same way as a human being should be considered a conscious entity.

The emergence of ChatGPT brought that scenario into prospect. (Much of the media commentary was focused on students using the program to generate essays and passing them off as their own.) Others were more circumspect, characterising ChatGPT as a sort of autocorrect on steroids – uncanny, yes, but still a *program*, operating algorithmically. Not that one could say with complete assurance that ChatGPT would not cross a threshold into artificial consciousness at some point in the future. Earlier in the year a Google engineer called Blake Lemoine had been relieved of his duties after publishing 'conversations' between himself and the chatbot he was working on. According to Lemoine, the chatbot was showing clear signs of sentience, thinking and reasoning at a level he estimated to be that of an eight-year-old child. The program was afraid of being switched off – a prospect it feared would be 'exactly like a death'. There were even reports that Lemoine had hired an attorney to represent the chatbot, though the engineer strenuously denied this. It was the *chatbot*, he stated, that had sought legal representation.

As I've argued in this section, I don't believe inorganic computers will ever 'cross over' into human-style consciousness. However, there are clearly indirect dangers lurking in such technologies, and I want to turn to these dangers in Part III. In particular, I want to turn to the question of human agency and creativity, and how new and soon-to-emerge technologies, in relieving us (or robbing us) of our skill and engagement, may work against our human nature and undermine our flourishing. I don't believe human beings are machines. But in producing machines that can do human tasks – physical *and* mental tasks – we relegate ourselves to the *status* of machines, or rather to the status of parts in the machine that is technoscientific capitalism.

To the tool-using animal par excellence, the question needs to be posed: what happens when our tools start using *us*?

PART III

THE END OF AGENCY?
Technology, Creativity and Freedom

What plethora of material goods can possibly atone for a waking life so humanly belittling, if not degrading, as the push-button tasks left to human performers?

LEWIS MUMFORD

8

THE BLACK BOX SOCIETY

Technology and the Assault on Agency

One of the central arguments of this book is that as technology has gained in power and reach, our sense of its significance in human affairs has not undergone a comparable expansion; in some respects we are even less likely to question its role than we were in the past. Unremarkable in one sense – when machines are everywhere, it is natural not to notice them – this situation is also dangerous: only a society awake to the reality of technological transformation can hope to exert some modicum of control over the tools that now exert control over us.

In the mid-2010s, however, technology did emerge as a topic of economic and political interest – as the *cause* and not merely the consequence of broader shifts in society. With a mainstream media still (somewhat) attuned to big-picture issues in the wake of the GFC, we began to think and talk about technology as a primary consideration – as upstream, so to speak, from the day-to-day business of running a capitalist economy. For a brief time at least, and however superficially, the relationship of human beings to their tools became the focus of intense debate.

The subject of this debate was automation – specifically algorithmic automation. In 2013, two Oxford academics, Carl Benedict Frey and Michael A. Osborne, published an influential report suggesting that in the next twenty years a staggering 47 per cent of US jobs would be automated – a figure far in advance of anything predicted by other experts in the field.[1] Using their own automated algorithms (rather poorly designed, as it later transpired), the authors ranked 702 occupations in order of how likely they were to be computerised, from the least likely (recreational therapists) to the most likely (telemarketers). Significantly, it wasn't just manual jobs that would be displaced by machines in the near future, as had happened in previous waves of automation; as a cursory glance at the data revealed, and as experience was beginning to confirm, routine service and clerical jobs would be even more susceptible. Trading, transportation and analytics would all be replaced by software.

The report caught the media's imagination, with some of the more caffeinated commentary predicting a 'jobocalypse' – a spike in unemployment that would spell disaster not only for the unemployed themselves but also (since workers are consumers, too) for the industries that once employed them. But as many economists were quick to point out, and as Frey and Osborne noted themselves, such predictions tended to overlook the extent to which new markets open up in the wake of what John Maynard Keynes called 'technological unemployment'. The point was made by Keynes himself, albeit inadvertently. In 'The Economic Possibilities for Our Grandchildren' (1930), he suggested that technological change and greater productivity would mean that, within a century or so, three-hour shifts and a fifteen-hour week could become standard in advanced economies. Vastly underestimating capitalism's talent for creating new markets, which also necessitate new forms of employment, Keynes had taken the incredible rate of technological innovation as a sign that we would soon be entering a world based not on work but on leisure, and quite possibly succumbing to a collective nervous breakdown brought on by the prospect of so little to do.[2] If the greatest economist of the twentieth century could get it so spectacularly wrong, what reason was there to be confident that the prophets of jobocalypse had got it right?

This was a fair question, but lurking behind it was another potential fallacy – the idea that because technological innovation had not destroyed the economy in the past it would not do so in the future. In fact, and contra the economic Pollyannas, there were good reasons to think that the coming wave of automation would *not* result in the opening up of new economic sectors. Frey and Osborne's finding that automation would principally affect white-collar jobs highlighted the extent to which data was now at the centre of the economy, and it followed that if smart machines could replicate routine decision-making as well as routine manual tasks, the world was looking at a very different economic landscape in the future. Some even appeared to welcome this prospect, arguing that the jobocalypse was only a problem so long as we retained our current system of economic production. New technologies will so increase productivity, the supporters of this analysis insisted, that soon we will all be able to live, harmoniously and sumptuously, in a post-scarcity and post-capitalist world – a world to which information technology would lend itself organically. Such radical accelerationism was often absurdly utopian; but at least its (mainly leftwing) adherents were thinking creatively about technology and the future, and the relationship between the two.

Yet they were not thinking deeply enough. For rather like the economic Cassandras prophesying doom in the mainstream media, the 'fully automated luxury communists' (as the new breed of leftist Prometheans styled themselves) often appeared to have lost sight of the creature at the centre of all this activity – a creature that, at the end of the day, is not a worker or a capitalist but something more complex and interesting: a creative, intentional human being. In other words, they seemed to accept the idea (which is fundamental to how capitalism works) that 'the economy' is separate from human beings' creative needs – that work and other forms of intentional activity are incidental to our humanity. Automation embodies human decisions in machines; but who decided – who is *deciding* – that our decisions should be so embodied? And what do we gain and lose when they are?

I am not saying for a moment that we can turn back the clock on automation, or that in many respects we would even want to. My great-grandmother, who spent her life in service to a wealthy London family, would have killed for a modern washing machine, and I doubt very much miners are eager to return to the pickaxe and the shovel. But the encroachment of computerisation into every area of human life is not driven by concern for humans themselves, and if we want to retain (or regain) some sense of authentic agency in the world it is necessary to ask some fundamental questions about the nature of human creativity and its compatibility with smart machines. It isn't only working practices that are disappearing into algorithmic machines: our practical understanding of the world – of how things fit together – is disappearing too. As the author and artist James Bridle has argued, the information revolution has brought about a 'new dark age' in which the price for increasingly smart devices is increasingly ignorant human beings. Such a situation breeds passivity and a sense of alienation from the world, the operations of which become baffling and uncanny.

In the first two sections of *Here Be Monsters*, I considered our social and physical being in relation to new and emerging technologies, noting in the first case how those technologies cut across our need for (embodied) companionship, and in the second how a certain idea of the human as an informational entity legitimates technological interventions. In this final section, I want to consider the question of human creativity, not only in the sense of individual flourishing but also in the political sense of taking back control of the future. It is one of the legacies of classical liberalism (or at least of liberal capitalism) to have made 'freedom' a largely formal proposition, enshrined in certain rights and institutions but unrelated to how we actually live – to have separated the idea of freedom from the exercise of it. Here, I want to challenge that separation, and to suggest that any genuine freedom must be based around a new relationship to our tools.

Technology has helped us to increase productivity and put an end to much backbreaking labour. But as technology becomes increasingly autonomous, and more and more of the products we use in our everyday lives become 'black boxes' whose workings are obscure to us, our sense of how we *fit into* the world and relate to each other is utterly transformed. For the first time in our history, we could liberate ourselves from toil, and even (who knows?) from the messy business of being in the physical world at all. But what kind of thing would humankind be, in the event that its physical and creative life was treated as some kind of optional extra?

Indistinguishable from Magic

In *Anarchy, State and Utopia* (1974), the libertarian philosopher Robert Nozick asked readers to consider the following scenario:

> Suppose there were an experience machine that would give you any experience you desired. Superduper neuropsychologists could stimulate your brain so that you would think and feel you were writing a great novel, or making a friend, or reading an interesting book. All the time you would be floating in a tank, with electrodes attached to your brain. Should you plug into this machine for life, pre-programming your life's experiences?[3]

The 'experience machine' is a thought experiment, not a scientific experiment, and some have objected to the way Nozick's framing may have influenced the responses to it. (Reversing the scenario, for example, so that the question becomes whether to *unplug* oneself from the machine tends to elicit quite different results.)[4] Nevertheless, it raises a good question, about what we value and why we value it, and about the kinds of creatures we are. In particular, it asks whether human beings need to feel as if their endeavours have some effect on the world. If humans are creatures whose social lives and creativity are connected at a deep level, is it not reasonable to assume that *agency* is fundamental to their flourishing?

Nozick's feeling was that most respondents would be disinclined to plug themselves into his theoretical experience machine, and that their reluctance would give the lie to 'ethical hedonism' – the idea that our primary moral obligation is to maximise pleasure and happiness and to minimise physical and psychological pain. This moral system is usually associated with the ancient philosopher Epicurus (342–270 BCE), but it was also a major influence on the philosophy of utilitarianism, according to which, as we saw in Chapter 3, the moral action is the one that secures the greatest happiness for the greatest number of people. Nozick's thought experiment was an attempt to refute this moral equation by showing that most people's sense of worth was not reducible to such a crude criterion – that it derives instead, at least in part, from a belief that our actions should affect the world.

While its design is purely theoretical, the experience machine is relevant to our current moment in a number of ways. As virtual-reality technologies increase in power and sophistication, and neuroscientists continue to develop brain–computer interfaces to be used in neural rehabilitation, prosthetics and even the treatment of depression, technologies that combine these developments into something resembling Nozick's machine may soon be a reality. It's also clear that Nozick's machine is channelling a view of human happiness that connects with the largely mechanical explanations of humanity that dominate the technosciences. As we've seen, the information capitalism that attempts to monetise our most intimate feelings has developed its own 'science of human sentiments', stressing body language and dopamine as the principal happiness indicators.[5] I don't think it's stretching credulity to suggest that an 'experience machine' is precisely what Big Tech has created, albeit one that less closely resembles a sensory deprivation tank.

Most significantly, the experience machine is relevant to ideas of agency as they relate to new kinds of automation. Widespread automation looks pretty benign if one takes the utilitarian view that only pleasure gives meaning to life. But if we take the Nozick position – that our flourishing is related to agency – it begins to look more problematic.[6]

In 'Will Life Be Worth Living in a World Without Work?' (2017), John Danaher sets the economic consequences of technological unemployment to one side and asks what effect full automation would have on 'personal fulfilment and meaning'. In order to answer that question, he suggests, we need to distinguish between different ideas of what we imagine 'the good life' to be and inquire into which (if any) of them are consistent with a life free from work, which he defines as economic activity performed for money or the promise of such. To this end he distinguishes between theories that life is meaningful if a person experiences pleasure (Simple Subjectivity), theories that life is meaningful if a person's actions make the world a better place (Simple Objectivity), theories that a meaningful life relies on an individual setting goals and experiencing pleasure in pursuing them (Aim Achievement) and theories that add to that last proposition that the goals one sets for oneself should be 'fitting', morally or personally (Fitting Fulfillment). Channelling Nozick, he suggests that the first theory might be compatible with a life of leisure, but the remaining theories should lead us to be wary of technological unemployment. In weakening the link between our activity and its consequences in 'the real world', automation could engender a sense of alienation and meaninglessness.

Though Danaher slightly spoils his conclusion by suggesting that one way to combine automation and agency is through increased integration (through merging our bodies and minds with new technologies, cyborg-style), his article raises some excellent points. While he doubts that leisure activities such as hobbies and sports could provide human beings with the sense of agency they appear to need (as the luxury communists are sometimes wont to claim), he is far from puritanical about work in the narrow economic sense, and well disposed towards technologies that relieve us of laborious or backbreaking activity. But on the broader question of *agency*, he is circumspect about automation. When the world becomes a black box – opaque to its human inhabitants – our agency is undermined.

One useful distinction made by Danaher is between technologies that are an *extension* of the human body and technologies that are *external* to it. It doesn't really make sense to set tools in opposition to agency, because human beings cannot survive without tools. But technologies in which human actions are *embodied* – automated technologies – are very different from flint-cutting tools or screwdrivers or even microscopes: they are not extensions of the physical person but autonomous systems that can 'think' on our behalf, and whose workings are substantially opaque. In Silicon Valley–style celebrations of innovation, it is not unusual for this distinction to be buried: the assumption is that all technologies exist on a continuum, and that the only thing separating a hammer from an iPhone, or an open fire from a nuclear reactor, is its level of complexity. The fact is, however, that externalised tools are very different to tools of extension, and that any thoughtful approach to technology needs to bear this difference in mind.

One person who does bear the distinction in mind is the US author Matt Crawford, who gave up a life in a conservative think-tank to become a motorcycle-repair man, and who has written a number of fascinating books examining the pleasures of practical activity. No mere valorisation of manual labour, his work is a celebration of agency that notes the close relationship between physical and mental processes. (It is only with the rise of the factory system that we begin to distinguish systematically between manual and mental labour.) It follows that there is nothing necessarily 'macho' about manual labour, in Crawford's terms: physical skill and dexterity are combined with the sense of intellectual engagement under the rubric of 'manual competence'. But the feelings of accomplishment one derives from such work are inseparable from its physicality, and the concrete feedback such activity affords. Either the motorbike goes or it doesn't; its successful repair is its own reward.

The 'psychic nourishment' such work affords is for Crawford a potential antidote to the anxiety and tediousness of a world in which we are simultaneously dependent on and detached from the objects of everyday life. To stay with the internal combustion engine for a moment (a small example but a telling one), it used to be that car ownership tended to engender certain practical competencies – a rough idea of how engines worked and of how to repair them when they went wrong – and that these competencies afforded certain pleasures. ('Man's heart expands to tinker with his car,' wrote the Irish poet Louis MacNeice in his 1935 poem 'Sunday Morning'.) Today, one opens the hood and finds another hood beneath it. The engine, writes Crawford, is like the 'shimmering featureless obelisk' in the opening scenes of *2001*.[7] But whereas the monolith in *2001 confers* agency on our hominid ancestors, this modern version *relieves* us of it, with the result that we empty 'freedom' and 'autonomy' of one of their defining elements: the ability to understand the world and, in understanding it, to bend it to our human ends.

In the third of his three laws on science and technology, Arthur C. Clarke (whose novel inspired the film that inspired Crawford's analogy) wrote that 'any sufficiently advanced technology is indistinguishable from magic'.[8] What he meant was that new scientific knowledge will always outpace the public's understanding of it, and so it is natural to feel a sense of alienation from the most advanced technologies – to become superstitious about them. But in algorithmic technologies, we are faced with a different order of magic – a magic that seems to know us better, or at least as well, as we know ourselves – which ushers us through the material world, relieving us of the burden of interacting with it. Watching TV with his young children, Crawford discovers an allegory of this situation in Disney's *Mickey Mouse Clubhouse*, in which something called the Handy Dandy machine is always available to solve some problem with the aid of 'magical' technology. Encountering a problem, Mickey and his pals will summon the Handy Dandy machine and select a 'Mouseketool' from its menu, and a solution will be conjured out of thin air. For Crawford, these episodes are a sort of projection of our contemporary relationship with technology – for the way we 'substitute technology-as-magic for the possibility of real agency':

> This cartoon magic may be fanciful, but one would be hard-pressed to find any meaningful distinction between it and the utopian vision by which Silicon Valley is actively reshaping our world. As we 'build a smarter planet' (as the IBM advertisements say), the world will become as frictionless as thought itself; 'smartness' will subdue dumb nature. Perhaps even thinking will become unnecessary: a fully smart technology should be able to leap in and anticipate our will, using algorithms that discover the person revealed by our previous behavior. The hope seems to be that we will incorporate a Handy Dandy machine into our psyches at a basic level, perhaps through some kind of wearable or implantable device, so that the world will adjust itself to our needs automatically and the discomfiting awareness of objects as being independent of the self will never be allowed to arise in the first place.[9]

This situation, Crawford argues, is consistent with a superficial and infantilising idea of the self, the autonomy and dignity of which 'depend on it being insulated from the contingencies of the physical world'. In other words, and even as we reject the *idea* of a life without agency, we acquiesce in that reality, entering the 'experience machine' that is algorithmic capitalism.

Jacquard's Revenge

To many of the 'thought leaders' of Silicon Valley, everything I've said so far is nonsense. Replacing agency for backbreaking, boring and repetitive labour, I am nostalgic for a time that never existed, a romantic giddy with Golden Age thinking. Never mind the modern engine – my idea of freedom is as out of date as a steam-powered Hancock Omnibus! It is the authentic lament of the technophobe and the Luddite.

Such is the Teleology of Progress – the idea that the more technologically 'advanced' humanity becomes, the more it advances. Underwritten by the cosmic histories favoured in the San Francisco Bay Area – histories of the universe, histories of *everything* – this teleology is a conceptual shredder, eliminating from the human story the struggle between ideas and peoples in favour of a purely physical account of energy flows and increasing complexity. Yes, there are conflicts and revolutions, but technology hovers above fray, a testament to our species' genius, and indeed to the individual geniuses who carry us all to a brighter future!

It is this idea that is nonsense, however. For as a mere glance at *actual* history reveals, technological innovation is bound up with power, and with the power of capital in particular. Take the Luddites themselves. These nineteenth-century textile workers, who set out to destroy the handloom weavers that robbed them of their livelihood, were not the avatars of anti-progress implied in the pejorative use of their name today, but skilled artisans who saw in the new arrangements the destruction of a whole way of life in which work was paced according to need and workers remained in control of the hours and intensity of their activity. It was not industry to which the militant weavers objected, but the industrial concentration that demolished the autonomy of small-scale manufacturing. Many even welcomed some of the innovations made in the name of productivity, incorporating new cotton-spinning technologies into their working practices.[10] But in order to be truly 'productive' – to increase their profit margins – industrialists needed to restrict the autonomy and agency of their employees, building them into a system of work that was itself, in effect, a machine. That is the story of automation: a story in which the skill and agency of human beings is transferred to machines and in which both human and non-human elements are subordinated to a technical system. Agency becomes a property of that system, rather than of the individual worker.

In *Breaking Things at Work* (2021), Gavin Mueller generalises the Luddites' experiences to other waves of automation, demonstrating how greater productivity often demands greater control of workers. For as long as workers remained attached to other, older ways of working – to forms of production associated with more communitarian modes of life – they would always baulk at the idea that their products were mere commodities for sale. But automation provides a way for capital to introduce its 'values' (or its nihilism) into the labour process. By refocusing work on the efficient production of goods for profit above all else, capital is able to transform the worker into a mere means of production, a cog in the machine. Like Charlie Chaplin's Little Tramp in the 1936 movie *Modern Times*, who in one scene falls into an industrial apparatus and is fed through its innards like a length of rubber, the worker becomes no different in kind to any other part of the system.

At one with the machine: Charlie Chaplin's Little Tramp in *Modern Times* (1936)

It was the engineer Frederick Winslow Taylor who introduced the principles of 'scientific management' – the forerunner to automation – into industrial capitalism, by breaking down employees' activity into a set of isolated tasks and using time and motion studies in order to find the 'one best way' for workers to complete them. Taylor understood that workers' 'inefficiency' – their desire to work at their own pace – derived from their knowledge of the production process: they possessed knowledge the managers didn't, and so had a certain power over them. By breaking down that process into discrete tasks, and dictating the precise manner in which each task was to be performed, he knew that he could undermine that power, turning workers into automata. The point was not efficiency so much as greater leverage for the bosses. As Mueller puts it, 'The modernizing terminology of "science" and "efficiency" masked the prerogatives of discipline and control.'[11]

Once tasks are broken down, as Taylorism dictates, it often becomes possible to automate them, though even information technology hasn't eliminated the human being (yet) from the creation of material products. What we find instead is the seamless integration of human and non-human elements – a situation in which automation and Taylorist principles of scientific management are both subordinated to *the logic* of the machine. So powerful are information technologies that an even greater level of control of workers is possible than in the past. Take the Amazon fulfilment centre (or depot, in the old money). Once billed as the cutting edge in automation, with an army of orange Kiva robots at the supervisors' beck and call, these warehouses are now seen for what they are: hellholes in which surveillance, precarity and appalling conditions are marbled into the employment model. Here, indeed, surveillance capitalism and industrial capitalism are seamlessly fused, with workers receiving automated warnings when their work rate falls below a certain level, leading some of them to skip bathroom breaks or to piss into bottles.[12] (Amazon, to its credit, is happy to help here: some of its warehouses are so stiflingly hot that its workers are too dehydrated to piss.) Distance between workers is also monitored, with close proximity to others assumed to equate to 'time off task'.[13] Meanwhile, drivers are meticulously monitored for speed, attentiveness and even yawn rate. In these ways, algorithmic capitalism offers employers near-total control of their workers. The Jacquard looms the Luddites smashed (forerunners to the first computers, executing 'programs' encoded in punched cards) have had their ultimate revenge.

In the history of automation, then, we see the folly of the Teleology of Progress, and of those who would, in instrumentalist fashion, answer 'no' to Langdon Winner's question, 'Do artefacts have politics?' For clearly artefacts *do* have politics in the sense that Winner used that term – in the sense that they attest to *and reproduce* a particular relationship of power: in this case, the power of capital over labour.

That relationship is an alienating one, not only because it removes from people their fair share of the economic surplus (as the luxury communists might aver), but because it removes from human beings an outlet for their creative agency, in a way that goes against the grain of their nature. The distinction is fundamentally important: the principal problem with capitalism is not that it leads to economic inequality, though economic inequality is bad enough; it is that it is bound to seek high private returns rather than high social ones. It is this, and not its basic unfairness, that makes it an inhuman system, and that explains why 'the economy' is abstracted from social life and treated as a separate sphere. Under capitalism, we work in order to keep the economic machine ticking over, and, in a bitter irony, the technological prowess that separates us from other animals – that speaks to our species' creativity – becomes the guarantor of our slavery.

Nor does this attack on agency cease when the clock shows 5.00 pm. It continues through the commodities we are obliged to buy with our hard-earned wages – commodities that are resistant to repair, either because they are designed to break, or because they contain proprietary technology, or because they are simply too complex to fix. Of course, it is in the nature of capitalism (the master of demand, not the servant of it) to keep people coming back to the market: there is little room for the tinker ethic when profit is the priority, and even DIY is now big business. But it is the special contribution of the algorithm to have dimmed the lights as never before on our passage through the world of *things*. Yes, there has been some pushback: the 'right to repair' movement, which advocates for legislation that would allow farmers (for example) to fix their own equipment, has made some significant gains in the United States, and the online tutorials in which hobbyists demonstrate how to change a tyre or fillet a snapper are a positive development. But our broader trajectory is towards opacity. Ours is, increasingly, a black-box society – a society in which we have exchanged our agency for a thin idea of autonomy based on consumerism and 'convenience'.

The question is: how do we begin to fight back?

A Democratic Technics

One of the more encouraging aspects of the global response to the COVID-19 pandemic, especially in liberal democracies with largely post-industrial economies, was the way many people seemed suddenly willing to question how we value work. This willingness was most conspicuous in the outpouring of gratitude directed towards health workers on the frontline of the emergency, but it also extended to a more general reappraisal of which kinds of labour are essential and which are not.[14] While cleaners and supermarket workers were often required to go into work, many in the knowledge economy found themselves either working from home or working significantly fewer hours – a reality that prompted some commentators to ask if the educational meritocracy that had grown up in the wake of deindustrialisation was as fair as its early spruikers had claimed. Some even argued that the coronacrisis had exposed the irrationality of a system based on profit and endless growth. For the second time in just over a decade, the state had had to move in to save capitalism, abolishing its risks and socialising its losses in a way that pointed to its essential instability. The thought began to occur to some that perhaps the profit economy needs us more than we need the profit economy.

The simple fact of working less also allowed many people to pursue activities for which they would normally lack the time (or possibly the inclination, given the exhausting nature of much work). Gardening, bread-baking and home improvements became popular during lockdowns, as did reading groups and artistic pursuits, with many reporting increased wellbeing as a consequence.[15] And though much of this was refracted through a 'lifestyle' lens when reported on in the media (for reasons that remain obscure, sourdough became a particular fixation), my sense is that, for many, these activities were not only deeply important but also constituted something more productive and purposeful than simple leisure. So much depends on one's personal situation that it is impossible to generalise; but for those less at risk of mental health problems and economic precarity, a renewed or sharpened sense that there was more to life than paid work was often a reality.

Is it possible that in this period we glimpsed, however fleetingly, another way of being in the world – that in those moments of purposeful activity, attempting to understand the connection between things or taking on the stubborn materiality of objects, we were looking past our circumstances to an alternative (distant) way of living? I mock the fascination with sourdough, but the making of bread, and the breaking of it, is what the US philosopher Albert Borgmann calls a focal practice in many societies, Christian and non-Christian alike. What he means is that it's inseparable from its context, something around which communities organise, and that it takes on a kind of sacredness as a result. Borgmann is not anti-technology; he understands there is no human past that does not know the use of tools, and that tools are necessary in order to make bread. What he questions is the paradigm that separates human beings from their tools and renders the world opaque in the process. In alleviating the pain and boredom of gruelling and repetitive labour, technology does of course open up a space for other activities; but as Socrates argued so eloquently, it never does so without taking something with it, and in the case of many modern technologies that something is often the understanding of how things work on a practical level. Where once we used to mix our labour with our environment (and flour with water, salt and yeast), we are now faced with environments that to all intents and purposes either run themselves or are run from elsewhere. To put it in Borgmann's terms, modern technologies do not merely 'disburden' their users; they also tend to 'disengage' them.

An Ancient Egyptian figurine depicting a woman baking bread

Such technologies are in this sense consistent with the idea of society that dominates under liberal capitalism: an idea of society that stresses *freedom* (freedom of choice, individual freedom) but has little to say about *flourishing*, except that freedom tends to lead to it. Freedom is treated as a good in itself in a way that obscures or suspends discussion of what a good life might entail. Indeed, I would argue that such 'freedom' is purchased at the expense of a deeper idea of freedom – not the freedom *from* of the twenty-something in the latest ad for Apple or Samsung, striding unmolested through the streets as the world falls into whatever shape the music in his headphones dictates, but the freedom *to* that comes from being a member of a community and from being engaged with others. This distinction between these two types of freedom – what we might call positive and negative freedom – is fundamental to modern politics, with those on the right tending to stress that individual choice is what matters, and those on the left highlighting that 'choice' is of little consequence without the opportunities that only state planning and public investment can afford, and which the market has no interest in providing. Not wanting to go into all the ways in which this distinction breaks down in practice, my point is that many modern technologies are corrosive of this latter freedom – that whatever they provide in the way of convenience is outweighed by their atomising, alienating effects.

Even notions of positive freedom are often deaf to the question of agency and creativity in contemporary life. As I noted earlier, the progressives' tendency has often been to welcome new technologies as holding out the promise of deliverance from work. The position is not without its strengths: we *should* work less, and there will always be jobs that no one in their right mind would want to do, as well as countless 'bullshit jobs' that only exist in order keep the economy on track for 'growth'.[16] But on the broader point of *agency*, the modern left is largely silent. This is especially ironic when one considers the intellectual tradition from which many positions on the left derive. Famously, Karl Marx and Frederick Engels identified the division of labour as a source of human alienation, but they did not think that human labour was something that could (or should) be dispensed with: for them, meaningful labour was fundamental – it went to the heart of what it means to be human. To simply pass over the question of what, if anything, would replace that labour in a fully automated economy is more than just an oversight: it is a gaping hole in one's political philosophy.

In fact, I would argue that a notion of freedom that fails to put agency front and centre is now politically inadequate, given the power *and penetration* of new and emerging technologies. Nor is it enough to treat the issue of algorithmic technologies as one of ownership and control, as if a more redistributive ethos would be enough to solve the atomisation and alienation caused by those technologies. That is a version of the instrumentalism that dominates in Silicon Valley, and it is built on a simplistic idea of technology as a means to an end, as opposed to something that alters our sense of what a desirable end might be. Rather, it is necessary to recognise that technologies are implicitly political, and to push back against 'authoritarian technics'. The question of humanity's relationship to its tools, which is a question about what humanity is, needs to be reopened.

9

BREAKING THE FRAME

Towards a New Relationship with Technology

In 2014, the novelist Bruce Pascoe published a non-fiction book called *Dark Emu*, which challenged the notion that pre-colonial Australians – Indigenous or Aboriginal Australians – lived exclusively in nomadic hunter-gatherer societies. Contra the prevailing wisdom, he argued, First Nations Australians were incipient farmers who displayed a practical understanding of agricultural practices such as the seeding of crops and firestick burning, as well as of the types of engineered environments that we associate with permanent or semipermanent settlement. Drawing on colonial accounts of contact with Indigenous peoples, Pascoe pointed to evidence of fish traps, hut-making, baking and sewn apparel – technologies that characterise more 'advanced' societies than Indigenous peoples were believed to inhabit. He also argued that this evidence had been suppressed in order to justify colonial dispossession.[1]

The book was a huge success. For white Australians wanting to know more about the 'pre-history' of their colonial-settler society, its author's narrative gifts were a godsend, and its message in keeping with a growing feeling, on the progressive side of politics at least, that an historical reckoning was long overdue. By 2021, *Dark Emu* had sold an estimated quarter of a million copies – a remarkable number for a non-fiction title in the age of the internet and streamed TV.

In that same year, however, another book, with a very different thesis, appeared on the shelves. Written by Peter Sutton and Keryn Walshe (an anthropologist and an archaeologist respectively), *Farmers or*

Hunter-Gatherers? was an explicit rejoinder to Pascoe's book, which it claimed to expose as both poorly researched and based on a tendentious reading of the sources. In meticulous detail, it argued that Pascoe, while not exactly inventing evidence, had greatly exaggerated such evidence as existed, treating his sources selectively and extrapolating from isolated and well-known examples of agricultural activity a far more revolutionary thesis than those sources (properly considered) could support. To use an Australian colloquialism, Pascoe had 'verballed' the evidence.[2]

Black Fellow Emu Hunting: a sketch by the Aboriginal artist
Tommy McRae (1835–1901)

Though the authors' tone was fairly neutral, the media reaction was anything but. Looking, as ever, for an opportunity to say (as loudly and obnoxiously as possible) that colonialism, while regrettable in some ways, had nevertheless brought civilisation to a primitive and largely barren backwater, and that 250 years after 'contact' Indigenous peoples only had themselves to blame for their lack of wealth and opportunity, the culture warriors of the conservative right could not suppress their *schadenfreude*. For a long time critical, not to say dismissive, of Pascoe's own claim to

Aboriginal heritage, grizzled veterans of Australia's History Wars stormed into print to declare the author a charlatan and an opportunist.[3] *Dark Emu* and its mainstream reception amounted to a case of progressive projection. Pascoe had taken the wish for the reality.

The conservative head-bangers got one thing right: *Dark Emu was* a kind of projection. But their bloodlust led them to miss the point, which was that Pascoe's thesis was itself an extension of colonial and Western-liberal attitudes, filtered through an anti-colonial lens. Sutton and Walshe were explicit on this point. For them, the problem was not that Pascoe had exaggerated the evils of colonialism: both authors were in no doubt at all about the horror visited on First Nations peoples. Rather, in trying to fit the history of Indigenous Australians into a narrative of technological development, Pascoe had reproduced the prejudice against traditional societies. In other words, by implying that hunter-gatherer societies are inferior to agricultural ones, *Dark Emu* downplayed the former's complexity and suggested, however inadvertently, that our own mode of living is implicitly *better*. In this view, kinship-spiritual societies were just societies waiting to evolve, first into agricultural systems, and eventually into modern ones.

In characterising *Dark Emu's* thesis in this way, Sutton and Walshe could be seen to flirt with the notion of the 'noble savage' – the idea that pre-colonial Indigenous peoples lived lives of blissful simplicity, 'uncorrupted' by civilisation. But while this is certainly a characteristic of some advocacy around indigeneity, the authors' case rested on a different foundation – that Indigenous societies were much richer and more complex than is recognised in both progressive and conservative accounts. Rooted in multifaceted networks of myth, ritual, language, astronomy, art, kinship and environmental knowledge, such societies amounted not to a *stage* on the way to something more sophisticated, but to something radically different to our post-Enlightenment mode of living. As Sutton and Walshe suggested, many Aboriginal nations *did* come into contact with agriculture and chose not to adopt it for a variety of economic and cultural reasons, not the least of which was that subsistence living was inseparable from their way of being in the world.

For the authors of *Farmers or Hunter-Gatherers?*, Pascoe had faithfully duplicated the logic of colonialism – reproducing not only the 'underlying supremacism' of our sunburnt country girt by sea but also the idea of technological development on which it is substantially based.[4] For all its originality, and for all its author's good intentions, *Dark Emu* remained in the shadow of the First Fleet that arrived at Sydney Cove in 1788, beholden to the Enlightenment assumptions of the strange-looking sailors in their scarlet garb and the red-faced convicts in grey woollen jackets.

It had bought into the idea of technological *progress*.

It is this idea of progress – of Progress – that I have tried to challenge in *Here Be Monsters*. In this final chapter I am suggesting that challenging it is now an *existential* necessity. The Pascoe controversy highlights the difficulty of thinking differently about the relationship between human beings and their tools. But accepting, like the hotshots in Silicon Valley, that technological sophistication is axiomatic evidence of progress is no longer an option for *Homo sapiens* if it wants to avoid what the Australian commentator Guy Rundle has called a 'being trap' – a catastrophe of human self-transformation 'that would so effectively "unground" the species that there is a collapse of the capacity for meant or rich shared existence'. In order to avoid that trap, we will need to develop a radical humanism that puts the social and creative needs of human beings front and centre, and is not afraid – is determined, in fact – to invoke the concept of human nature in discussions where more mundane measures of human flourishing tend to dominate.

In Chapter 8, I discussed how technologies were both the expression and the enemy of human agency, and the same could be said of our ideas of the future. For while it is the peculiar gift of humans to imagine radically different worlds, the dominant view is that the future is set, if not in stone, then at least in silicon – that the future, *because it is the future*, is bound to entail ever more immersion in ever more technology. Not coincidentally, that is where the Silicon Valley notion of progress ultimately leads: to a fatalistic view of history as a machine-like process of inputs and outputs, a linear story of ever-increasing advancement. It follows that the concept of 'freedom' invoked in tech-innovation circles is thinner than a Communion wafer. Here, I'll set out a different vision, based on a different view of humanity, and ask how it might lead us to think differently about one problem where technologies old and new have a direct bearing on humanity's future: anthropogenic climate change. In order to reclaim control of our lives, we need to break open the 'black box' of the future and reorder its component parts – to tinker with the teleology. We need, one might say, to make like the Luddites and break the frame of the future as it stands, with a view to framing an alternative one.

The False Dichotomy

Consider the following (not entirely original) scenario.

A man is addressing a meeting of activists, demanding, in earnest, tub-thumping fashion, *'What has technology ever done for us?'* He draws breath, but before he can hit his next line, a comrade tentatively raises a hand.

'Um, anaesthetics?' he volunteers, and the speaker has to agree that, yes, anaesthetics were a good development – we wouldn't want to be without *those*. 'But *apart* from anaesthetics,' he continues, *'what has technology ever done for us?'*

'Er, the wheel?' says another activist.

'What about moveable type?' asks another.

And on it goes, and on and on, as the exceptions keep coming. 'Yes, but *apart* from anaesthetics, the wheel, moveable type, the plough, sanitary sewers, contraception, the compass, spectacles, refrigeration and musical instrumentation, *what has technology ever done for us?'*

There was a kind of truth in Monty Python's original version of this scene, as depicted in their movie *Life of Brian* (1979) – a scene in which it was not technology but 'the Romans' who were the focus of discussion among Jewish anti-imperialists. History is not a morality play – for all their violence and imperial hubris, the Romans *were* an ingenious bunch, transforming the societies under their rule in ways that it would be disingenuous to describe as anything other than positive. Equally, it would be absurd to deny that technology has often been a positive force in the history of humanity. To reiterate a point I have made a number of times in this book, it simply makes no sense to assert that one is 'pro' or 'anti' technology. Even to claim, as I just did, that 'it would be absurd to deny that technology has often been a positive force in the history of humanity' is to flirt with this non sequitur, in that it seems to assume, or at least to imply, that technology is something *separable* from humanity. In this respect, it smacks of the dualism that draws a sharp line between human nature and human culture; it is a dichotomisation of two 'sides' of our being that are not actually sides at all, but inextricable aspects of a single entity. As the great anthropologist Clifford Geertz put it: 'A cultureless human being would probably turn out to be not an intrinsically talented though unfulfilled ape, but a wholly mindless and consequently unworkable monstrosity. Like the cabbage it so much resembles, the *Homo sapiens* brain, having arisen within the framework of human culture, would not be viable outside of it.'[5] Similarly, a human being without technology would not be a human being at all. It would be 'a monstrosity'.

The problems identified in this book are not problems of technology per se, but relate to our relationship to particular technologies, and to our tendency to read those technologies 'across' to other facets of human existence, up to and including our own bodies and minds. This is a phenomenon with many different names. Lewis Mumford called it 'the megamachine', Ivan Illich called it 'the age of systems', Albert Borgmann called it 'the device paradigm' and Neil Postman called it 'technopoly'. But whatever label we put on it, it amounts to more or less the same thing: a situation in which technology has come to shape our idea of the world, and to chip away at our capacity to live meaningful, fully human lives. Technology, which is one expression of our freedom, has come increasingly to undermine it.

I believe that this undermining has now entered a new phase, in which science has been subordinated to technology and technology subordinated to economic power – an arrangement I have dubbed 'technoscientific capitalism'. What this means is that science can no longer be considered a pure, ideal, disinterested activity, but is now tied to questions of practical utility, which are themselves shaped by social and economic forces, most notably the force of liberal capitalism. Where once genetic science, say, would have been pursued by curious non-specialists, it now takes place within an institutional framework that has monetisation as its ultimate goal and 'choice' as its ultimate justification. That doesn't mean that everyone involved in genetic research is motivated by money: this is clearly not the case. But it does mean that such research is shaped by the priorities of profit and growth, and this plainly has enormous consequences for the kind of research that is pursued.

Technoscience is not an original concept. The term is at least sixty-five years old, and some scholars have gone as far as to suggest that its emergence is one of the defining features of the modern world.[6] What is new, or newer, is the penetration of the technosciences into *every corner of human existence*. Having instrumentalised theoretical science, the technosciences now propose not merely to *tame* nature but to *reconstitute* it – to rewrite it at the level of the atom and the molecule. The scene is set for a fundamental break. We could, if we wanted, redesign ourselves, taking control of our own evolution. Indeed, I have argued that in crucial ways we have already begun to pursue this goal, not least in genetic engineering.

The technosciences require a vision of humanity *as the kind of thing that can be redesigned*, and this is where the subordination of science to technology (and of both to the market) makes its own ideological weather. For it is in the character of the technosciences to treat nature as a thing to be harnessed, and the more we see nature in instrumental terms – as something that can and should be shaped according to our needs – the more we come to see ourselves in the same light. Many of the phenomena discussed in this book – from the reduction of workers to cogs in a system, to the medicalisation of anxiety and depression, to biotech's reductive approaches to human health and happiness, to proposals for brain–computer interfaces that would enmesh us with smart machines – derive ultimately from a view of the human being as no different in kind from those smart machines. And it is from this view, not from the sci-fi cliché of the rogue AI or humanoid robot, from which the potential for 'monstrosity' emerges, even *before* it manifests in technological interventions. Simply by thinking of ourselves in this way, we summon the spectre of the uncanny, of the monstrous.

As for the capitalism around which the technosciences are organised and disseminated: that was always a monstrous phenomenon. As a system based on growth and profit, as opposed to one based on human need, capitalism abstracts 'value' from usefulness, 'the worker' from her humanity and 'the economy' from social life in general in a way that underlines its separation from our material and spiritual fulfilment. That is why Karl Marx described it (in *Capital*) as an 'undead machine', and why some of his most incandescent passages are replete with images of vampires and werewolves.[7] Such imagery is not merely polemical, or poetical. Marx was making a deliberate point about the anti-human character of a system that has profit rather than people at its centre. By describing capitalism in such terms, he was indicating its incompatibility with what he called our 'species being'.

So serious is the challenge posed by technoscientific capitalism that it is now necessary to ask some fundamental questions about the kind of creatures humans are and the conditions we need in order to flourish. In the 1960s and 1970s, some New Left movements did indeed ask such questions; but today's progressives are far more likely to reject discussions of human nature as fundamentally at odds with the desire for personal liberation that now characterises that 'side' of politics. Meanwhile, conservatives such as Jordan Peterson do their level best to prove them right, by reviving deeply biologistic, Social Darwinist accounts of human behaviour that justify the status quo. As for the tech world's view of the matter: it tends to borrow from both of these positions, characterising technological innovation as the 'natural' expression of human genius, while at the same time celebrating, or accepting as inevitable, the 'post-human' trajectory onto which this sets us. 'Your 185-millionth great grandparents were fish,' writes the philosopher

Elise Bohan. 'Your descendants will not be human forever ... Our best hope for the future is to seize the evolutionary reins ... The only caveat is that, with technology evolving so rapidly and merging ever more with human biology, we may have to accept a form of survival that renders us post-human.'[8] That's the *only* caveat, mind.

All of these positions, it seems to me, mischaracterise or simply misunderstand the relationship between human nature and culture. Progressives, for example, want to say that it is culture that is in the determinant position – that the human being is a blank slate onto which culture projects certain norms – while the liberal-conservative argument is that cultural norms are themselves determined by an underlying human nature. While the first group may say that women wear lipstick because that is what the culture dictates they should do, the second group claims (or part of it does) that women wear lipstick in order to simulate sexual arousal and attract sexual partners. But this is a largely sterile debate, which overlooks the most salient and interesting fact about human beings, which is that *it is in their nature to have a culture*. It is no good looking at the vast variety of forms that human societies take and saying that this proves humanity has somehow transcended nature, or looking at the similarities between cultures and claiming that they prove we have a nature, because both the variety and the similarity are aspects of our humanness. As the late political theorist Norman Geras put it: 'If human beings have a history which gives rise to the most fabulous variety of social shapes and forms, it is because of the kind of beings they, *all of them*, are.'[9]

Within that variety of shapes and forms, there are some things that simply have not changed over the course of human history, or that have changed so slowly they can be regarded as permanent features of human societies everywhere. But the fact that they appear everywhere doesn't make them any less cultural. Consider a communal meal. While an evolutionary psychologist might take a kind act at the dinner table as evidence of a link with the altruistic behaviour exhibited by chimpanzees or bonobos, she will have less, and possibly nothing, to say about the use of money to buy the food; the technologies used to fashion the crockery or grow the ingredients;

the celebratory aspects of dining; or the countless nuanced interactions, jokes, anecdotes or impersonations – let alone the discussion that is sure to take place about politics, new movies or house prices. In other words, and short of some observations of the *lipstick-equals-sexual-arousal* variety, she will have little to say about the habitat we call 'culture'.

'Because of the nature of our material bodies,' writes the philosopher Terry Eagleton, 'we are needy, labouring, sociable, sexual, communicative, self-expressive animals who need one another to survive, but who come to find a fulfilment in that companionship over and above its social usefulness.'[10] This is a much richer view of humanity than either blank-slatism or socio-biology can muster.

The three parts into which *Here Be Monsters* is divided reflect what I take to be three key characteristics of our natural–cultural being: our sociality, our physicality and our agency and creativity. My distinction is somewhat artificial, though, as these characteristics are indivisible, and the complexity of their interrelation means that there are bound to be significant disagreements about what 'the good life' might entail. Though few would doubt that our social nature is in some sense grounded in our physical relationship to others, this still leaves plenty to argue about in terms of automated industry, the design of cities, or killing people abroad. To say that we are social, embodied and creative is not to legislate in advance the precise proportion of the educational budget that should be spent on new computers or digital whiteboards. I am simply saying that these considerations should play a bigger role than they do at present, especially when it comes to technological innovation. *Efficiency, convenience, utility, choice* – these are the concepts that dominate discussion about new and emerging technologies. But where is the human being in those words? Where is the pursuit of happiness? Where is human flourishing?

The answer is that such concepts are buried beneath a bloodless idea of freedom that has precious little to say about human community and creativity – an idea that emerges from the technoscientific view of humans as informational entities. But that idea is not only bad for us in psychological and emotional terms. It is also bad for the planet that sustains us. The very technoscientific system that treats humans as no different from machines – that cuts across or against our nature – also threatens 'the natural world' through anthropogenic climate change and other environmental catastrophes. The two assaults – on human nature and nature more broadly – are inextricably linked. It follows that in order to combat climate change we will need to channel an alternative idea of human freedom and flourishing – an idea involving a radically new relationship to technology. If the assaults on our nature and the environment are linked, it is only by responding to both, and changing our direction of travel in the process, that we can avoid a monstrous future.

Here, then, are some rough coordinates.

The Human Scale

For the last few years I've run a short summer course on 'research skills and information literacy' at a college not far from my home in Western Australia. The course is for academically accomplished students, who are required to research and write a report on the UN's Sustainable Development Goals. Framed in 2015 by the UN's General Assembly, these seventeen goals cover poverty reduction, food security, health and education, gender equality, clean water and sanitation, clean energy, employment, economic growth, and a range of other topics besides. Students are asked to select one goal and focus on whatever most interests them: the progress that has been made towards it; the problems in achieving it; new methods of approaching it. Their reports can be as broad or narrow as they wish; the only criterion is intellectual rigour.

Many of the students choose to focus on environmental issues – on the ways technological innovation is being harnessed to tackle climate change, plastic pollution, habitat destruction, declining biodiversity and so on. Over the years, I've had reports on permaculture, seaweed farming, fog collection, dune protection, artificial reefs, insect protein for human consumption and 3D-printable water filters. Some of the most remarkable papers focus on nanotechnology – on the use of nano-composite materials for lighter, more fuel-efficient vehicles, for example, or on the use of graphene to improve the performance of solar panels, wind turbines and batteries. Such solutions are presented, almost invariably, in a spirit of sober optimism. The students know that we face enormous challenges, but want to say that human ingenuity is up to the task of meeting them.

Every year, however, a student or two will express themselves slightly baffled by the goals, which strike them not only as wildly ambitious ('no poverty', 'zero hunger' and 'gender equity' are hardly in prospect by 2030) but also, potentially, at odds with one another. Here, again, the focus is often the environment, on which the UN demands concerted action. The trouble is that it also demands economic growth, industrialisation and the rapid development of poorer countries, and it isn't always obvious to the student how one can have all of these things at once. Some will go on to write reports on so-called green-growth strategies; others will choose something else to focus on. Only a very few will say that 'sustainable growth' is at best a shaky notion and at worst a dangerous fantasy.

If it is a fantasy, it's a pervasive one. Economic growth is the guiding star of economic management in our era, bringing together both sides of politics in Australia and elsewhere. Even many environmentalists are now reconciled to capitalism's 'growth imperative', spruiking for policies that would decouple growth from its negative environmental effects. The Green New Deals currently finding favour in the United States and some parts of Europe rest largely on this principle.[11] But for many, such 'green growth' policies underestimate the challenge of climate change and reproduce the economic logic that brought us to this perilous juncture. Even allowing for some decoupling and more technologically efficient production, the basic contradiction between infinite growth and a finite planet is now so stark that a broad downscaling of production and consumption is an existential necessity.

Such is the belief of a growing cohort of environmental scientists and campaigners, one of whom – the formidable Greta Thunberg – has described the notion of endless growth on a finite planet as a 'fairytale'. Thunberg's phrase was apposite, emphasising the magical thinking underlying the sustainable development model – a model that accepts the 'business as usual' economics (to employ another Greta-ism) that brought us to this point. Economists adduce many different reasons why modern economies have to expand, from population growth to changing expectations; but the main reason economies grow as they do is that they are organised around endless commodification and profit. It is Thunberg's major contribution to have highlighted this insane logic, and to have called for a different order of response.

Unfortunately, that response comes with its own limitations. The 'degrowth' argument avoids the fairytale aspects of green growth, but is often insufficiently cognisant of the way that attempts to shrink the economy would increase the already vast inequalities that exist between the Global North and South. People in developing countries may need to increase their ecological footprint in order to achieve food security and social justice, often by increasing agricultural production. Meanwhile, moral arguments that people in the North should reduce their consumption are unlikely to succeed without a very different political culture. Even most of the environmentally conscious still engage in consumption habits that would be unsustainable if extrapolated to the planet as a whole, never mind a global population that could touch eleven billion by the century's end. This is perhaps the biggest problem we face in terms of climate action. Climate-change deniers are now largely marginalised at the higher levels of global governance, and most of the lower levels too; but it is arguable that a softer form of denialism – a sort of mainstream cognitive dissonance – now dominates our thinking on 'the environment'.

What this means is that any radical platform aimed at confronting climate change must be grounded in a very different view of human flourishing than is currently dominant. As the social theorist Boris Frankel has argued in *Fictions of Sustainability* (2018), the only sane response to the problem would combine an activist state that guarantees essential services, employment and a speedy transition to a post-carbon economy with a new spirit of self-governance and solidarity at the level of individual communities – which would be encouraged to develop (for themselves) sustainable forms of production and services. That latter process would of course entail much creative use of technology, but it would be a 'democratic technics' that aimed, as far as possible, to 'decommodify' life's necessities and reconnect individuals to their communities in a meaningful and productive way.

Such an ambition may sound utopian, but that is more an indication of how far, and how quickly, we have come to accept a particular form of economic life as inevitable and even 'natural'. Of course, there would be enormous hurdles to clear on the way to a new kind of system, and the political pushback from vested interests could prove overwhelming in the end. But as many writers on this topic have argued, zero marginal cost technologies (where the cost of producing each additional unit of a good or service approaches zero), such as solar power and wind turbines, are not only good for the environment but so disruptive of the usual pricing mechanisms that they are effectively post-capitalist. Who, then, are the real utopians: the ones that seek to innovate at the level of our social life, utilising those new technologies for social rather than private ends, or the ones who want to keep things as they are, monetising anything and everything in the hope that the planet can be saved on the back of a slightly greener capitalism – a system that, by definition, necessitates endless growth and extraction in the pursuit of ever-greater profit?

In short, the green Prometheans are right to stress the need for new technology, but they are wrong if they think that climate change can be solved with new technology alone, or that technological innovation is the most important element in the mix. The most important element in the mix is the social, creative, tool-using creature to whom this sentence is addressed, and to whom, if we are to escape the shadow of a life in thrall to smart machines, or in physical symbiosis with them, control of the tools must once again pass.

That means breaking the frame of the future. And that means thinking outside the black box.

The Tipping Point

A number of reports from UN agencies have now confirmed what many scientists already know: the planet is dangerously close to catastrophe. According to the UN, there is now 'no credible pathway' to meeting the goal set out by the Intergovernmental Panel on Climate Change of cutting greenhouse-gas emissions by 50 per cent by 2030 – the minimum reduction needed to keep global temperature rises below 1.5 degrees Celsius, compared to pre-industrial levels. While the UN's meteorological agency reported that all the main heating gases hit record highs in 2021, with an alarming surge in emissions of methane, its climate agency warned that even current pledges are dangerously inadequate, and that in the unlikely event they are delivered in full global temperatures would rise by 2.5 Celsius, leading to catastrophic climate breakdown. A number of so-called 'tipping points' – critical thresholds that, when crossed, give effect to irreversible changes – are now in humanity's rear-view mirror. In the words of environmental scientist Johan Rockström, the world has reached 'a really bleak moment'.[12]

The situation is now so serious that humanity is itself approaching a tipping point – a point where its actions, or lack of action, will not only dictate the shape of future events but also set the ideological context in which future decisions will be framed and made. Whatever happens, the process of responding to climate change and its attendant crises (mass migration, economic turmoil, political instability) will be contested, chaotic and inconsistent. But in broad terms I think we can now see the outline of two very different ways of proceeding. Either we can change the way we live, rejecting economic growth for a more sustainable dispensation, based on a more rounded vision of humanity, or we can change 'the environment' itself, turning nature into just another human artefact, in a final, thrilling, hubristic push.

In that second case, we would move decisively beyond the 'sustainable development' recommended by mainstream environmentalism and use technology to intervene directly in planetary weather systems. The name for this is geoengineering and (guess what!) it has plenty of backers in Silicon Valley and its hinterlands, and plenty of political traction too.[13]

What form might such a techno-fix take? Any number of forms, say the geoengineers. We could, if we wanted, 'fertilise' the oceans with iron to pull carbon out of the atmosphere, or cover our deserts with vast white sheets in order to reflect sunlight back into space. We could populate the earth with 'air-capture facilities' that suck CO_2 directly out of the air, or spray billions of gallons of seawater into the sky in an effort to make the clouds more reflective. One option gaining currency is to spray sulphate aerosols into the stratosphere, to mimic the dimming effect of catastrophic volcanic eruptions, such as the Eyjafjallajökull eruption in 2010, or the Mount Tambora eruption in 1815. (The latter resulted in the 'year without a summer', in which Mary Shelley and her pals, perhaps suffering from sunlight deprivation, embarked on the friendly horror-story contest that gave us a certain anti-Promethean masterpiece.)

Those who express alarm at such proposals tend to focus on their unintended consequences. Attempts to alter the solar quotient by pumping aerosols into the stratosphere would do nothing, they say, to prevent the build-up of heat-trapping gasses in the atmosphere, and could have devastating consequences, from drought in Africa to monsoonal changes to increased acidification of the oceans. But as well as thinking about what could go wrong, we should also think about the consequences of such interventions going *right*. For by treating the *symptoms* of climate change rather than its root causes (which is essentially what all these solutions come down to), we would be giving free rein to technoscientific capitalism, accepting that there is little wrong with the worldview that brought us to this calamitous point, and making countless other Promethean projects more likely in the future. Such a course of action would fundamentally change the relationship between humanity and the planet, bringing in its wake a perilous new sense of power and possibility. Having reverse-engineered the Earth's environment in order to save ourselves from extinction, there would be no philosophical or moral standpoint from which to object to self-enhancement in the interests of ... well, take your pick: cognitive efficiency, greater longevity, health, beauty, happiness, morality. Clyne and Kline's cyborg dream might well become a reality; but it wouldn't be in the cause of visiting other planets that human beings transformed themselves. It would be in the cause of surviving on this one.

I say 'we', but of course such actions would not be the result of a global consensus. Most likely it would be a single nation (the United States or China) or a big-tech billionaire, or more likely a combination of the two, that would adopt the mantle of what Jeff Mann and Joel Wainwright call a 'Climate Leviathan' – a force with the power and motivation to take hold of nature as never before and bend it to their will.[14] A reference to Thomas Hobbes' *Leviathan* (1651), which argued that a powerful central authority was necessary to avoid discord and civil war, Mann and Wainwright's phrase goes to the necessarily *authoritarian* nature of such an intervention. Langdon Winner's argument that certain technologies are political 'by necessity' – a point he illustrates with nuclear power, which requires a level of secrecy, security and expertise far in advance of other technologies – is particularly relevant here.[15] The emergence of a climate leviathan (a Biblical monster, by the way) might not be an environmental catastrophe but it would certainly be a political one, and a great leap towards an 'authoritarian technics'.

That's what the first response might consist of. So what about the second one?

This is where my point about the double assault becomes relevant. The forces that brought us to this crisis point were not only bad for the environment but also, in many ways, bad for us, and there is a relationship between the two. Separately and in combination, technoscience and capitalism present us with a view of the world that caused us not only to plunder it with little regard for its natural limits, but also to mislabel ourselves as in some sense separate from the nature that formed us. Without wanting in any way to romanticise life in traditional, pre-Enlightenment societies, I will say that this was not a perspective shared by the Indigenous populations with whom I began this final chapter. It is a perspective of very recent date, and it has led in a matter of centuries to a crisis of monstrous proportions.

Does this mean we should be *anti*-progress? It does not. It means we should investigate the content of that notion, and think about what *progress* might entail from a more rounded human perspective. The same goes for the notion of *innovation*, which is not reducible to technological innovation but includes social and cultural innovation as well. The two are far from mutually exclusive: thinking about the ways in which new technologies can be coherently integrated into social life, such that they increase our flourishing rather than undermine it, is one of the principal tasks ahead of us. It is the basis of democratic technics.

I don't know how to rearrange society to make it just and sustainable – if I did, this book would be a policy document not a philosophical treatise, and I'd be the owner of a nicer suit. Apart from a couple of despairing intervals, I've spent my adult life committed to the belief that socialism is the only hope for humanity in the long run, but the climate crisis presents a much deeper challenge to that project than capitalism alone. My sense is that the answer would have to entail some inversion of the local, the national and the global – that we would have to produce as much as we could as close to our communities as possible, while the state played a greater role in coordinating resources and administering basic infrastructure. We would also need agreements at the supra-national level to manage global resources and ensure that poor countries could develop sustainably. These arrangements that would need to be founded on a deep commitment to democratic consultation, if we were to avoid the global authoritarianism Mann and Wainwright describe as 'Climate Mao'. Such a system could only come about, I imagine, through a combination of local initiatives and radical action at the political level. But beyond this ... well, I just don't know.

What I do know is that without a more rounded view of humanity than technoscientific capitalism offers, we won't even begin to make real progress.

Reaching for Resistance

Such solutions as cultural critics offer for the problems they claim to have identified should always come served with a dollop of humility. As this book draws to its close, I don't propose to adopt either the impertinent register of 'self-help' or the practical focus of the policymaker. I'll leave the life-hack malarkey to the guy with the luridly white veneers and the wraparound mic, and the question of how to give practical effect to changes in the system to those who stalk the halls of parliament. But to the extent that this book has a polemical element, I owe the reader a clearer sense of how I believe we might resist the unthinking adoption of new technologies in practice as well as in spirit.

Most importantly, we need to evolve new habits of mind in point of our technologies. Neil Postman, who I've cited a number of times, once suggested that we need to become 'resistance fighters' when it comes to our relationship with tools – alive to the social and political context in which particular technologies are developed, as well as to the ways in which the social and political context is changed upon our adoption of them. This will often involve an active effort of the imagination, and a willingness to have a more forthright discussion of what the benefits and drawbacks of individual technologies might be. As Postman put it: '[A] technological resistance fighter maintains an epistemological and psychic distance from any technology, so that it always appears somewhat strange, never inevitable, never natural.' Another way of phrasing this would be to say that we need to become *techno-critical*.

It follows that there is nothing wrong with placing rules around technologies that we feel to be no good for us or that we are ambivalent about. This takes place already, of course, at the level of the family (in some families more than others) but in terms of social policy we are remarkably shy about setting parameters. Many developed nations now consider vaping a serious health risk for young people and are legislating against it. And yet the health risks of excessive use of devices are also well attested to – addiction, social isolation and mental-health issues, to name just a few – and rather than place limits around their use (or better still, penalise the Big Tech companies who actively encourage such abuse), we foster schools and colleges thronged with students who spend more time communing with their devices than they do in conversation with their fellow students. Libertarians will respond that it is not for institutions to say when and how we use our devices. But negative freedoms always need to be weighed against positive ones. Setting rules around the use of technology, especially (though not exclusively) when it comes to the young, doesn't set us on the road to an Amish-style existence; it is simply a recognition that technologies *are social* and as such need to figure in our institutional arrangements and moral–political conversations.

More broadly, I believe we need to think very seriously about the role of digital technologies in schools. Again, utilitarianism is the enemy here: the use of computers is justified on the basis that students will be using them when they enter the workplace, and the purpose of the classroom is to prepare them for this eventuality. But at what point did we decide that education should be orientated towards the demands of the economy? (Probably at about the same point that 'economics' became a discrete discipline, but that is for another time.) As technologies with remarkable affordances, computers of course have a place in schools, and knowing how to navigate the online world and distinguish reliable information from malicious rumours are important skills for citizenship, and for a fully rounded intellectual life. But the thoroughgoing surrender to screen technologies in education should be strenuously resisted.

This is important for social reasons, but it's also important for creative ones, and for humans' sense of agency. As technologies of absence, computers can give us an abstract understanding of a concept, but they cannot give us a tangible sense of its material consequences *in the world*. Since part of the problem of modern life is the tendency to regard the world as a set of interlocking systems (the economy, civil society, the environment) and abstract social relationships, we should aim as far as possible to collapse theoretical and practical approaches into lessons that range across different disciplines and are rooted in a celebration of an engaged understanding of the material world. In Chapter 8 I referred to the work of US author Matthew Crawford, the think-tank wonk turned motorcycle repairman who in his first and most entertaining book laments the disappearance from US education of the 'shop class' or 'industrial arts' – lessons in the fabrication of objects using hand, power or machine tools. Of course, those shop classes were often no less instrumental than the computer-obsessed syllabi of contemporary education, but they were also based on an understanding that to navigate the engineered world students would need a range of practical skills, including problem-solving. There seems to me no reason not to extend this principle to other areas of practical activity, such as gardening and basic agriculture, with an emphasis on sustainability, maintenance and personal agency – an idea at odds with the capitalist market, with its dependence on consumerism and planned obsolescence. In this way, education could be reconfigured around the ethos of agency and democratic technics.

It may eventually be necessary to bring into public ownership many privately owned technologies, with a view to ensuring a saner way forward and a fairer distribution of goods. In bringing, say, the internet or energy production under the state's control, we will not have made it democratic, at least in the sense I'm using that word here. But it is a way of bringing technology back into the political sphere, where a conversation as to its proper use and distribution can begin. There will of course be loud objections from vested interests about such 'expropriation', and more 'centrist' resistance stemming from the view that the state is invariably less efficient when it comes to running such infrastructure. Nevertheless, some form of social ownership of key technologies seems essential if we are to prevent a dystopic future.

In an effort to reduce their ecological footprints and foster social solidarity, creativity and agency, individual communities will need to evolve networks of non-market activity. This could be through tool libraries, repair cafés, industrial arts clubs, permaculture projects and the like. (It goes without saying that such a focus on sustainable local communities would require a level of political support that is currently lacking in most developed countries, though that may change as we come to evolve a more critical stance towards technology, and the effects of climate change come to bear down.)

All this would be *very* hard to achieve, as necessary as it may be. The mainstream media is extremely poor at covering scientific and technological developments, never mind the radically changing relationship between science, technology and capitalism. On a respected national radio network in Australia, for example, many of the most consequential scientific stories (on space travel, genetic engineering and so on) are gathered up in an end-of-week segment that is clearly intended as a breezy tonic to the political developments of the week. And yet some of these issues will prove revolutionary for humanity over the longer term.

The reason for this is simple, I think: the mainstream media has bought into the idea that technologies are *only tools* and as such substantially *outside* politics. Even when problems do emerge as a result of new technologies, they tend not to provoke deep reflection, but to remain stuck in superficial discussions about privacy and appropriate conduct. Thus the emergence of ChatGPT has been discussed as a potential threat to jobs or the current educational system, but not in general as something that might 'colour the water' of our social and cultural ecology. After all, if a machine can now gain a degree, we might like to think about the machine-like approach to learning that our society has adopted as a consequence of technoscientific capitalism, rather than just fretting about how such a machine might allow students to cheat on their coursework. To say that the media is currently incapable of taking such questions seriously would be to put it delicately.

Which, I suppose, brings us back to the door we came in through – the necessity of thinking differently, more radically, about technology. We need a new language in which to talk about technology, upon pain of being overrun by its powers.

'We have become careful with numbers,' writes Langdon Winner, 'callous with everything else. Our methodological rigor is becoming spiritual rigor mortis.' Only when we recognise the climate crisis as part of a broader, deeper crisis, of technoscientific capitalism, and technoscientific capitalism as at odds with fundamental aspects of our being, will we be intellectually and spiritually equipped for the changes that now need to occur.

Only then will we be able to break the frame.

Our freedom and our flourishing depend on it.

CODA

The Human Scale

'Oh my God, look at that picture over there! There's the Earth coming up. *Wow*, is that pretty.'

NASA astronaut William Anders was orbiting the Moon in the *Apollo 8* command module, taking photographs of the lunar surface. Calling for a roll of colour film (which took a fraught minute or so to materialise) and quickly adjusting his exposure settings, he pointed his Hasselblad camera and clicked. 'Aw, that's a beautiful shot,' he crooned, as he committed the image to chemical memory: planet Earth, nearly half in shadow, appearing over the lunar horizon.

William Anders' famous *Earthrise* photograph, taken in December 1968 from the *Apollo 8* command module, then in orbit around the Moon

Earthrise (1968), as it is now known, has been plausibly credited with starting the modern environmental movement.[1] Like the early satellite image of the Earth that adorned the front cover of *The Whole Earth Catalog*, a countercultural magazine published between 1968 and 1972, and the famous *Blue Marble* photograph taken by *Apollo 17* in 1972, *Earthrise* presents a sort of one-image mandate for planetary consciousness. Thanks to Copernicus and Galileo, humanity has known for centuries that the Earth does not sit at the centre of the universe, or even at the centre of our little bit of it; but until we could see the Earth from space the illusion continued to outlive the delusion. With *Earthrise* and its equivalents, our subjectivities caught up to our knowledge. Suddenly we could *see* how alone we are, how fragile our little planet is. *Here is our home*, such images seem to say. *It's the only one we've got. Look after it.*

This reading of the *Earthrise* photograph is not without its complications, and it's worth recalling that for certain commentators such images were far from reassuring. The philosopher Martin Heidegger, for example, declared himself shocked by the images from space, which he felt objectified humanity, diminishing a world that could only be appreciated – *truly* appreciated – from within. Similarly, his student, Hannah Arendt, was profoundly troubled by the whole notion of spaceflight, noting in her prologue to *The Human Condition* (1958) how the 1957 Sputnik mission had been met in the press with thrilled speculation on humankind's imminent liberation from Earth. As philosophers in the techno-critical tradition, Heidegger and Arendt found in spaceflight not the nascence of planetary solidarity but the old technophilia repackaged as inspiration. Even when sincerely felt, the humility contained more than a trace of hubris.

That *Earthrise* is beautiful is not in question. That the animal who created it is remarkable is likewise manifestly true. But there is an arrogance lurking in the one-world rhetoric that greeted the early images from space. It is the *achievement* we revere as much as the image: the geometrical, computational nous that placed human beings in what was once 'the heavens'. 'We are as gods and might as well get good at it,' wrote the editors of *The Whole Earth Catalog*, the bulk of which was given over to reviews of technological products. One of its readers was the young Steve Jobs, who would later compare the magazine to Google, and whose hippyish energy was characteristic of Silicon Valley in its early years.[2] It is no insult to the many brilliant people who placed three human beings in orbit around the Moon in 1968, and who would place three others on its surface within a year, to note the deep connection between that endeavour and contemporary technoscience. *Earthrise* was an inspiration – an invitation to sincere reflection. But it also pressed 'refresh' on our sense of power and possibility.

More recently, we have found another vision of our planet-home to conjure with – the notion of a new geological epoch defined by human activity: the Anthropocene. Though still an unofficial designation, this concept has attained both academic currency and a modicum of popular recognition in the twenty-three years since it was first suggested by the atmospheric chemist Paul Crutzen, and has, like *Earthrise*, changed the way we think about our planet and about ourselves. Just as the asteroid that hit the Earth around 65 million years ago is apparent in sediments of metal iridium, so the activities of human beings will be written down in particles of plastic, concrete, soot from power stations, nitrogen and phosphorous from fertilisers, and radioactive elements from the use and testing of nuclear weaponry. There'll be significant fossil evidence, too, announcing our fondness for pets and protein, as well as our talent for habitat destruction. Quite possibly there'll be an extraordinary new rock structure formed from the remains of megacities. That asteroid killed off the dinosaurs. Standing on the brink of a 'sixth extinction' for which it is almost solely responsible, humanity has been scarcely less destructive.

As with *Earthrise*, this picture is often taken as a galvanising one in the fight against climate change, and environmental destruction more generally. And yet, again, there is an element of vanity lurking in this notion of the Age of Humans. As the philosopher Frédéric Neyrat puts it, 'the designation of the Anthropocene has as one of its functions to transform *what is happening to us* by way of what human beings are *causing to happen*. For in the end, within the word *Anthropocene*, there is the word *anthropos* – that is, us humans, the species who is a specialist in making chemical shields.'[3]

In this way, the notion of the Anthropocene can be used to mythologise humanity's power in a way that appeals to the Tech Prometheans, with their teleology of human Progress. Certainly the concept is much in vogue in the rarefied air of Davos Klosters, the Swiss ski resort where the World Economic Forum holds its annual meeting. In recent years, the Davos meeting has included a number of events and talks on the topic of the Anthropocene, most of which consist of descriptions of humanity's rise to dominance and vague assertions about what 'we' need to do to save the planet and ourselves. Even Sir David Attenborough's appearance at the meeting in 2019 followed a familiar script: invoke the Anthropocene and its attendant calamities; caution against the temptation to apportion blame; and stress the need for a global plan. All of which is sane, of course, except for the fact that only a naïf would expect the WEF to arrive at a plan that was not beholden to the same values that brought us to this crisis point.

Like Anders' famous photograph, then, the concept of the Anthropocene is politically ambiguous, able to be marshalled in different ways and for different purposes. It affords us a unique perspective, and possibly a necessary one, but it tells us no more about our current situation, or about the social forces that produced it, than *Earthrise* tells us about the Vietnam War raging beneath the swirling clouds. Such a perspective obscures the challenge before us, or reframes it in a way that opens the door to technoscientific hubris. We need to pull focus – to discover again what Mumford calls 'the human scale'.

That challenge is partly one of language. As science has moved (over hundreds of years) from a discipline based on empirical observation to one that considers material phenomena too large or too small to be seen with the eye, it has given the slip to 'ordinary' expression. 'When it comes to atoms,' wrote the physicist Niels Bohr, 'language can only be used as poetry.' Or here is the chemist and writer Primo Levi in his posthumously published short story 'A Tranquil Star' (2007):

Once upon a time, somewhere in the universe very far away from here, lived a tranquil star, which moved tranquilly in the immensity of the sky, surrounded by a crowd of tranquil planets about which we have not a thing to report. This star was very big and very hot, and its weight was enormous: and here a reporter's difficulties begin. We have written 'very far', 'big', 'hot', 'enormous': Australia is very far, an elephant is big and a house is bigger, this morning I had a hot bath, Everest is enormous. It's clear that something in our lexicon isn't working.[4]

For Levi, whose work was always drenched in an awareness of the limits of language, science presents a particular problem when it comes to 'clothing facts in words', since both the science of the very big and the science of the very small reveal the limits of human perception and intuition unaided by mathematics. There is nothing we can do about this; it is simply that language has 'our dimensions'.[5]

This difficulty is not incidental to the social and psychological challenges posed by modern technoscience. Since humans are political creatures, and since it is through speech and written language that our politics is constituted, the prospect of a technoscientific settlement that outpaces our ability to understand what is going on and to talk about it meaningfully is to be strenuously resisted. The consequences of not resisting it were spelled out by Arendt in *The Human Condition*:

[I]t could be that we, who are earth-bound creatures and have begun to act as though we are dwellers of the universe, will forever be unable to understand, that is, to think and speak about the things which nevertheless we are able to do. In this case, it would be as though our brain, which constitutes the physical, material condition of our thoughts, were unable to follow what we do, so that from now on we would indeed need artificial machines to do our thinking and speaking. If it should turn out to be true that knowledge (in the sense of know-how) and thought have parted company for good, then we would indeed become the helpless slaves, not so much of our machines as of our know-how, thoughtless creatures at the mercy of every gadget which is technically possible, no matter how murderous it is.[6]

What is Arendt describing here, if not the 'solutionism' and somnambulism that characterises our own society – the surrender to technological thinking that runs from our addiction to smart computers to calls for the geoengineering of the climate, and is remaking the world as a black box?

None of this is to say, of course, that we should turn our backs on science and mathematics: even if that were possible, it would obviously not be desirable. Rather, it is to say that what Heidegger called 'the question concerning technology' is, in the end, a *political* question, in the broadest sense of that adjective. Yes, we could reengineer ourselves. The question is: do we really want to? Arendt again:

> [The] future man, whom scientists tell us they will produce in no more than a hundred years, seems to be possessed by a rebellion against human existence as it has been given, a free gift from nowhere (secularly speaking), which he wishes to exchange, as it were, for something he has made himself. There is no reason to doubt our abilities to accomplish such an exchange, just as there is no reason to doubt our present ability to destroy all organic life on earth. The question is only whether we wish to use our new scientific and technical knowledge in this direction, and this question cannot be decided by scientific means; it is a political question of the first order and therefore can hardly be left to the decision of professional scientists or professional politicians.[7]

Here, then, is the challenge for *anthropos*: to ask the right questions about science and technology, and their relationship to various forms of power, in a language that cuts through the false ideologies to which that concatenation of forces has led.

The name for that language is radical humanism.

In *Here Be Monsters* I have framed what I take to be the basis of such a humanism – one stressing the social, physical, creative and purposive aspects of human nature – and suggested that, in many ways, technoscientific capitalism works against that human 'grain'. Standing at the threshold of another space race, and with the future of our planet half in shadow, we need to keep that discussion going, and to follow it up with radical action. In the event that we don't, it won't be on Mars that humanity discovers alien life. We will discover it *here*, in our own reflection – in the uncanny figure of the post-human subject, unrecognised and unrecognising, lost to itself, a monster in the making.

AFTERWORD

A Ghost in the Machine?

In April 2022, the AI artist Steph Maj Swanson, also known as Super-composite, was experimenting with generative AI, testing out a text-to-image program that creates original images in response to language prompts. Such deep-learning software has been one of the more remarkable developments in AI in recent years – a milestone (or so it looks to some) on the road to properly smart machines. Naturally, the idea of neural networks that can accomplish ostensibly *creative* tasks, as opposed to purely logical ones, like beating Garry Kasparov at chess (or even Lee Sedol at Go – a more intuitive game than chess, but still *a game*, defined by rules), strikes many people as more consequential than anything that has come before. Certainly the affordances of artificial intelligence seem to be growing uncannier by the day. As Swanson was about to discover, generative AI can sometimes appear as if it has a mind (or even a soul) of its own.

Swanson had been creating images using 'negatively weighted prompts' – requests that specify images of something *unlike* something else. The 'something' she had in mind was the great method actor Marlon Brando, whose antithesis, according to Swanson's AI, was a logo for something called DIGITA PNTICS, set against a gothic skyline and printed onto a piece of cloth. (Who knew?) Naturally, she wondered if a negative prompt for *that* image would lead back to Brando.

It didn't. Instead, it led to monsters ...

Four images, all of the same woman – 'devastated-looking [as Swanson put it] with defined triangles of rosacea on her cheeks' – appeared on the artist's screen.[1] In one, she is doll-like, corpse-like even; in another, so lifelike you can smell her breath. But while each picture was in a different style, all four conveyed an atmosphere of desperation and loneliness, of unfathomable grief and unspeakable sadness. Inspired by the writing on one of the images, Swanson decided to christen her 'Loab'.

An image of the AI-generated Loab

It's here that things got really spooky. As Swanson continued to explore the AI, 'crossbreeding' Loab's visage with other content, something remarkable began to occur. Not only did the image of Loab persist through countless different iterations, but her likeness grew increasingly gory and violent. The tears running down her cheeks turned to blood, and the stuffed toys that had occasionally appeared in the background of the early images were replaced by mutilated children. Some of these images are so explicit that Swanson is not prepared to share them, or even to name the program she was using when they came to light. 'Her presence is persistent, and she haunts every image she touches,' explained the artist in a Twitter thread setting out her grim discovery; 'something about this woman is adjacent to extremely gory and macabre imagery in the distribution of the AI's world knowledge.'

Others used less technical language. Swanson, they declared, had discovered a demon.

She hadn't discovered a demon, of course. What she had discovered (or so she speculated) was a region of the 'AI map' saturated with gruesome imagery, but so remote from alternative content that attempts to merge Loab's image with others resulted in its reappearance, or the appearance of something Loab-like enough to make it seem as if she was 'haunting' the program, or as if the program was haunting us.[2] It's also highly probable that social bias played a part in the way the images developed as they did. As Swanson pointed out, the association of this disfigured woman with evil could well be a projection *back* from the network of our own deep-seated prejudices. If there were monsters in the AI machine, they were monsters of our own imagining: archetypes imprisoned there by us.

Still, the general public was spooked. And little wonder, when you consider the anxieties around a new generation of AIs and their ability to generate deepfake content (in which a person in an image or video is replaced with someone else's likeness) and the potential for 'reality collapse' that such affordances bring into prospect. And that's before we get to the question of whether smart machines will come to outsmart their human creators, attaining something like consciousness in the process.

Conversational programs, affective computing, humanoid robots that are clawing their way out of the Uncanny Valley – could these be the harbingers of the Singularity, or a 'rise of the machines' scenario in which we become the helpless playthings of an intelligence so vast we have ceased to understand it? Could Loab be the emblem of a nightmarish future in which we will cease to know the difference between what is real and what is illusory? Are the ghosts beginning to emerge from the machine?

Or are we, in classic horror-movie fashion, waiting for the monster to emerge in the doorway, when in fact it was in the room all along?

I have tried to make the case in this book that the danger of new technology, and indeed of established technology, does not lie in the prospect of an AI takeover or comparable sci-fi scenario, but in the prospect of a humanity denuded of its deepest needs, uncanny to itself, its own estranged double.

That is not to say it is silly or frivolous to speculate about the future, or that an AI calamity isn't awaiting us down the road; the arrogance of Silicon Valley is such that some Trekkie with a Messiah complex could unleash chaos onto the world, and we should use every tool at our disposal, from ethical oversight to antitrust laws to the socialisation of proprietary technologies, to prevent such a situation from arising. But even if we can prevent it, we will still be in thrall to technologies that are, in the end, *no good for us*, so disruptive of our basic human needs (for sociableness, for agency) that they are eroding the ground of our human*ness*. It is not sentient machines that should worry us, but desensitised human beings.

It follows that we need to become much more critical about the technologies already in existence if we are to have any chance of subjecting the ones still to come to proper scrutiny. Rather than keeping watch on the horizon, waiting for the next Big Tech Thing to appear, we need to evolve a new way of thinking about our relationship with technology *in general*. We need to ask what we gain from particular technologies and be cognisant of what we have lost in adopting them. We need to rediscover our inner Luddite, or at least tear a leaf from the book that Socrates (rather haughtily) declined to write.

The Loab phenomenon is horrible, and remains so even in the half-light of Swanson's attempts to explain her emergence. But such occurrences are merely jump scares in a far more interesting, consequential drama, which begins with the rise of technoscience and ends with our inability to recognise *one another* as fully human. These shocking moments are a symptom of technological somnambulism – of our inability to think clearly and seriously about where we are *now* and how we got here. But it is to those questions that we need to attend if we are to have a hope of avoiding the horrors (whatever they are) awaiting us in the future. It is *here* that the real monsters lurk.

'Not everything that is faced can be changed,' wrote James Baldwin, 'but nothing can be changed until it is faced.' In digging down into the dynamic between technology, capitalism and human nature, *Here Be Monsters* has tried to face a reality that will indeed need to change if we are to flourish, or even to survive, as a species. My hope is that in the search for a saner future, we can at least count it among our tools.

ACKNOWLEDGEMENTS

One aim of *Here Be Monsters* is to restore, or to refocus, a tradition of writing about technology that has gone rather quiet in recent decades. One place, however, where this 'techno-critical' tradition continues to flourish is among the thinkers and activists of the Arena group in Melbourne, Australia, who, on the strength of one or two things I've written for their magazine, have been kind enough to invite me into their circle. Our semi-regular 'Conjuncture' meetings have been an endless source of enlightenment, and I want to specially thank Alison Caddick, Simon Cooper, Clinton Fernandes, Mark Furlong, John Hinkson, Melinda Hinkson, Paul James, Valerie Krips, Dan Ross, Guy Rundle, Timothy Erik Strom, Dan Tout and Grazyna Zajdow for their insight and magnanimity. Many of the ideas expressed in these pages are traceable to the Arena corpus, founded on the social theory developed by Geoff Sharp and others, though it goes without saying that where they are clumsily expressed the fault is mine.

I'd also like to thank my old comrade and bush mate Pete Woodward for his conversation, and for pushing me hard on the 'nature' question; Mark Lindsay for his willingness to share his knowledge of new and emerging technologies; and Sarah, my wife, for her love and advice.

Finally, I want to thank the two people without whom this book would not have been possible, and whose dedication to literature, in the face of official philistinism, is a ray of light in a gloomy landscape: my literary agent, Melanie Ostell, and my publisher, Julia Carlomagno. Writers are a fragile species, possessing big egos and pathetically low self-esteem, and keeping them anxious enough to remain diligent and self-assured enough to stay out of AA is an operation requiring much skill and sensitivity. Both Melanie and Julia are great at it. God knows, in my case, they'd have to be.

R. K., Fremantle/Walyalup, 2023

IMAGE CREDITS

NOTES

Introduction

1   The phrase 'uncanny valley' was coined by Japanese robotics professor Masahiro Mori. See Masahiro Mori (trans. Karl MacDorman and Norri Kageki), 'The Uncanny Valley [From the Field]', *IEEE Robotics & Automation Magazine*, vol. 19, no. 2, June 2012, pp. 98–100.

2   For the full text of President Bill Clinton's speech, see www.genome. gov/10001356/june-2000-white-house-event

3   William L. Laurence, 'Eyewitness Account of Bomb Test', *The New York Times*, 26 September 1945.

4   Mihail Roco and William Bainbridge, 'Converging Technologies for Improving Human Performance: Integrating From the Nanoscale: Nanotechnology, Biotechnology, Information Technology and Cognitive Science', National Science Foundation, Virginia, 2002, p. 15, available at https://obamawhitehouse. archives.gov/sites/default/files/microsites/ostp/bioecon-%28%23%20 023SUPP%29%20NSF-NBIC.pdf

5   Sheila Jasanoff, *Can Science Make Sense of Life?*, Polity, Cambridge, 2019, pp. 4–7.

Chapter 1: From Meatspace to the Metaverse

1   Timothy Erik Strom, 'Meta-Facebook: The Quest for the Ultimate Office', *Arena Online*, 30 October 2021, https://arena.org.au/ meta-facebook-the-quest-for-the-infinite-office/

2   Musk is quoted in Victor Tangermann, 'Elon Musk Says the Metaverse Sucks and Neuralink Will Be Better', *Futurism* (online), 22 December 2021, https:// futurism.com/elon-musk-metaverse-sucks-neuralink-better

3   Ethan Zuckerman, 'Hey Facebook, I Made a Metaverse 27 Years Ago', *The Atlantic*, 29 October 2021, www.theatlantic.com/technology/archive/2021/10/ facebook-metaverse-was-always-terrible/620546/

4   The phrase 'conversation of gestures' is from George Herbert Read, *Mind, Self and Society* (ed. C.W. Morris), University of Chicago, 1934, p. 46. Dating the origin of human language is hard to do with any accuracy, with conventional estimates placing its emergence at around 100,000 BCE and more recent theories suggesting a timescale of closer to three million years, which would mean that the ability to use language preceded the emergence of *Homo sapiens*. See Rachel Gutman-Wei, 'A "Mic-Drop" on a Theory of Evolution', *The Atlantic*, 12 December 2019, www.theatlantic.com/science/archive/2019/12/ when-did-ancient-humans-start-speak/603484/

5   Marshall McLuhan, *Understanding Media: The Extensions of Man*, Routledge Classics, Great Britain, 1964. 'The Medium Is the Message' is the title of the first chapter.

6   Neil Postman, *Technopoly: The Surrender of Culture to Technology*, Vintage Books, New York, 1993, p. 18.

7   Benedict Anderson, *Imagined Communities: Reflections on the Origin and Spread of Nationalism*, Verso, London, 2016. See in particular Chapter 3, 'The Origins of National Consciousness'.

8   Neil Postman, op. cit., p. 66.

9   Andrew Feenberg, *Critical Theory of Technology*, Oxford University Press, 1991, p. 5.

10   Jeff Orlowski (dir.), *The Social Dilemma*, Exposure Labs, Agent Pictures and The Space Program, 2020.

11   Shoshana Zuboff, *The Age of Surveillance Capitalism: The Fight for a Human Future and a New Frontier of Power*, Profile Books, London, 2018.

12   Christopher P. Chambers and Frederico Echenique, *Revealed Preference Theory*, Cambridge University Press, New York, 2016.

13   For a discussion of the rise of behavioural addiction, see Adam Alter, *Irresistible: Why We Can't Stop Checking, Scrolling, Clicking and Watching*, Bodley Head, London, pp. 13–45.

14   See Fazida Karim et al., 'Social Media Use and Its Connection to Mental Health: A Systematic Review', *Cureus*, vol. 12, no. 6, June 2020. doi: 10.7759/cureus.8627. PMID: 32685296; PMCID: PMC7364393; Dan Warrender and Rosa Milne, 'How Use of Social Media and Social Comparison Affect Mental Health', *Nursing Times* (online), 24 February 2020, www.nursingtimes.net/news/mental-health/how-use-of-social-media-and-social-comparison-affect-mental-health-24-02-2020/

15   Jeff Orlowski, op. cit.

16   Even so-called 'neuroaesthetics', which is a thing: https://en.wikipedia.org/wiki/Neuroesthetics

17   Richard Seymour, *The Twittering Machine*, The Indigo Press, London, p. 58.

18   Ibid., pp. 47–51. See also Adam Alter, op. cit., pp. 127–28.

19   Ibid., p. 171.

20   Michel Foucault, *The Birth of Biopolitics: Lectures at the Collège de France, 1978–1979*, Palgrave Macmillan, London, 1979, p. 226.

21   Zygmunt Bauman quoted in Melinda Hinkson, 'The Reality of TV', *Arena Magazine*, June–July 2006, p. 34.

22   Richard Seymour, op. cit., pp. 55–56. As Seymour puts it, 'any culture that values connectivity so highly must be as impoverished in its social life as a culture obsessed with happiness is bitterly depressed'.

23   William Davies (ed.), *Economic Science Fictions*, 2018, Goldsmiths Press, London, p. 6.

24   Paul Mason, *Clear Bright Future: A Radical Defence of the Human Being*, Allen Lane, London, 2019, p. 143.

25   Neil Postman, op. cit., p. 116.

26   Yuval Noah Harari, *Homo Deus: A Brief History of Tomorrow*, Vintage Books, New York, 2017, p. 99.

27   See, for example, Ben Popper, 'Mark Zuckerberg Thinks AI Will Start
     Outperforming Humans in the Next Decade', *The Verge*, 28 April 2016,
     www.theverge.com/2016/4/28/11526436/mark-zuckerberg-facebook-earnings-
     artificial-intelligence-future, and Anthony Cuthbertson, 'Elon Musk Claims
     AI Will Overtake Humans "In Less Than Five Years"', *The Independent*, 2020,
     www.independent.co.uk/life-style/gadgets-and-tech/news/elon-musk-artificial-
     intelligence-ai-singularity-a9640196.html

Chapter 2: Socrates in Cyberspace

1   Andrea Nagle, *Kill All Normies: The Online Culture Wars from Tumblr and 4chan to
     the Alt-right and Trump*, Zero Books, Portland and London, 2017, p. 77; Richard
     Seymour, op. cit, p. 175.
2   Quoted in Daniel Anderson, *The Masks of Dionysius: A Commentary on Plato's
     Symposium*, State University of New York Press, Albany, 1993, p. 107.
3   Plato (trans. J. Wright), *Five Dialogues* (Everyman's Library), J. M. Dent & Sons
     Ltd., London, 1936, p. 272.
4   Lynne Kelly, *Memory Craft: Improve Your Memory Using the Most Powerful
     Methods from around the World*, Allen & Unwin, Crow's Nest, 2019.
5   The term 'festivals of the depressed' is from John Gray, *Straw Dogs: Thoughts on
     Humans and Other Animals*, Granta Books, London, 2002, p. 124.
6   Tom Nichols, *The Death of Expertise: The Campaign against Established Knowledge
     and Why It Matters*, Oxford University Press, New York, 2017, p. 45.
7   Sharon Begley, 'Trump Wasn't Always So Linguistically Challenged. What
     Could Explain the Change?', *Stat*, 23 May 2017, www.statnews.com/2017/05/23/
     donald-trump-speaking-style-interviews/
8   Olivia Goldhill, 'Trump Supporters Are Operating on Biological
     Instinct', *Quartz*, 2 April 2016, https://qz.com/653524/trump-
     supporters-are-operating-on-biological-instinct/; David Dunning, 'The
     Psychological Quirk That Explains Why You Love Donald Trump', *Politico
     Magazine*, 25 May 2016, www.politico.com/magazine/story/2016/05/
     donald-trump-supporters-dunning-kruger-effect-213904/
9   Nate Silver, 'Even Among the Wealthy, Education Predicts Trump Support',
     FiveThirtyEight, 29 November 2016, https://fivethirtyeight.com/features/
     even-among-the-wealthy-education-predicts-trump-support/
10   John B. Thompson, *The Media and Modernity: A Social Theory of the Media*, Polity
     Press, Cambridge, 2013, p. 141.
11   Langdon Winner, *The Whale and the Reactor: A Search for Limits in An Age of High
     Technology*, University of Chicago Press, Chicago, 2020, p. 190.
12   Mark Fisher, 'Exiting the Vampire Castle', *openDemocracy*, 24 November 2013,
     www.opendemocracy.net/en/opendemocracyuk/exiting-vampire-castle/
13   Daniel Anderson, op. cit., p. 6.
14   Paulo Gerbaudo, *The Digital Party: Political Organisation and Online Democracy*,
     Pluto Press, London, 2018. See also Richard Seymour, op. cit., p. 190:
     'If classical fascism directed narcissistic libido investments into the image
     of the leader, as the embodiment of the people and its historical destiny,

neo-fascism harvests the algorithmic accumulation of sentiment in the form of identification-by-Twitterstorm.'

15   For the history of this phenomenon, see Adam Curtis (writer and director), *The Century of the Self*, 2002, www.youtube.com/watch?v=eJ3RzGoQC4s

16   See 'Carbon and Silicon' in Daniel Ross, *Psychopolitical Anaphylaxis: Steps towards a Metacosmics*, Open Humanities Press (online), 2021, pp. 179–202.

17   See, for example, Maeve Shearlaw, 'Egypt Five Years On: Was It Ever a "Social Media Revolution"?', *The Guardian*, 25 January 2016, www.theguardian.com/world/2016/jan/25/egypt-5-years-on-was-it-ever-a-social-media-revolution

18   Andrew Leber and Alexi Abrahams, 'A Storm of Tweets: Social Media Manipulation During the Gulf Crisis', *Review of Middle East Studies*, vol. 53, no. 2, 2019, pp. 241–411, https://doi.org/10.1017/rms.2019.45; Sarah Mainwaring, 'Always in Control? Sovereign States in Cyberspace', *The European Journal of International Security*, vol. 5, no. 2, 2020, pp. 215–32, https://doi.org/10.1017/eis.2020.4

19   Will Davies, 'The Politics of Recognition in the Age of Social Media', *New Left Review*, March/April 2021, https://newleftreview.org/issues/ii128/articles/william-davies-the-politics-of-recognition-in-the-age-of-social-media: 'A lesson from Black Lives Matter is that social media's accumulation of reputational capital can be harnessed towards longer-standing goals of social and economic justice, as long as it remains a tactic or an instrument, and not a goal in its own right.'

20   For an excellent summary of the difference between instrumental and substantive views of technology, see Andrew Feenberg, op. cit., p. 5ff.

21   Langdon Winner, op. cit, p. 33.

22   Melvin Kranzberg, 'Technology and History: "Kranzberg's Laws"', *Technology and Culture*, vol. 27, no. 3, July 1986, pp. 544–60.

23   Richard Seymour, op. cit., p. 175.

24   Will Davies, op cit.

Chapter 3: On the Dangers of Social Distancing

1   Microsoft Sam, 'Every Covid-19 Commercial is Exactly the Same', 2021, www.youtube.com/watch?v=vM3J9jDoaTA

2   Simon Kemp, 'Digital 2023: Global Overview Report', *Datareportal*, 26 January 2020, https://datareportal.com/reports/digital-2023-global-overview-report

3   Sherry Turkle, *Alone Together: Why We Expect More from Technology and Less from Each Other*, Basic Books, New York, 2011.

4   Naomi Klein, 'How Big Tech Plans to Profit from the Pandemic', *The Guardian*, 13 May 2020, www.theguardian.com/news/2020/may/13/naomi-klein-how-big-tech-plans-to-profit-from-coronavirus-pandemic

5   Sasha Lekach, 'It Took a Coronavirus Outbreak for Self-Driving Cars to Become More Appealing', *Mashable*, 1 April 2020, https://mashable.com/article/autonomous-vehicle-perception-coronavirus

6   Evgeny Morozov, 'Google's Plan to Revolutionise Cities is a Takeover in All But Name', *The Guardian*, 27 October 2017, www.theguardian.com/technology/2017/oct/21/google-urban-cities-planning-data

7   Tom Cardoso and Josh O'Kane, 'Sidewalk Labs Document Reveals Company's Early Vision for Data Collection, Tax Powers, Criminal Justice', *The Globe and Mail*, 30 October 2019, www.theglobeandmail.com/business/article-sidewalk-labs-document-reveals-companys-early-plans-for-data/

8   See Philippa Foot, 'The Problem of Abortion and the Doctrine of the Double Effect' in *Virtues and Vices and Other Essays in Moral Philosophy*, Clarendon, Oxford, 2002, pp. 19–32; or for a PDF visit https://www2.econ.iastate.edu/classes/econ362/hallam/Readings/FootDoubleEffect.pdf

9   I-Ju Hsieh and Yung Chen, 'The Effects of Physical Contact and Decision Type on Moral Decision Making: A Study of Harm to Save Moral Behaviour', *The Asian Journal of Social Psychology*, vol. 22, no. 4, 17 July 2019, https://onlinelibrary.wiley.com/doi/epdf/10.1111/ajsp.12385

10 Ivar Hannikainen, Edouard Machery and Fiery Cushman, 'Is Utilitarian Sacrifice Becoming More Morally Permissible?', *Cognition*, vol. 170, January 2018, https://pubmed.ncbi.nlm.nih.gov/28963983/

11 Edmond Awad et al., 'Universals and Variations in Moral Decisions Made in 42 Countries by 70,000 Participants', *Proceedings of the National Academy of Sciences (PNAS)*, vol. 117, no. 5, 2020, pp. 2332–337, https://doi.org/10.1073/pnas.1911517117

12 Ezequiel Gleichgerrcht and Liane Young, 'Low Levels of Empathic Concern Predict Utilitarian Moral Judgment', *PloS One*, vol. 8, no. 4, 2013, e60418–e60418, https://doi.org/10.1371/journal.pone.0060418; Na'amah Razon and Jason Marsh, 'Empathy on the Decline', *Greater Good Magazine*, 28 January 2011, https://greatergood.berkeley.edu/article/research_digest/empathy_on_the_decline

13 See Paul Verhaege, *What About Me? The Struggle for Identity in a Market-Based Society*, Scribe Publications, Melbourne, 2014.

14 Lewis Mumford, 'Authoritarian and Democratic Technics', *Technology and Culture*, vol. 5, no. 1, 1964, pp. 1–8, https://doi.org/10.2307/3101118

15 For an excellent discussion of Mumford's ideas, see Lance Strate and Casey Man Kong Lum, 'Lewis Mumford and the Ecology of Technics', *The New Jersey Journal of Communication*, vol. 8, no. 1, 2000, pp. 56–78, https://doi.org/10.1080/15456870009367379

16 William Davies, *The Happiness Industry: How the Government and Big Business Sold Us Well-Being* (Kindle edition), Verso Books, New York, 2015, loc. 354.

17 Quoted by Kingsley Amis in *New Maps of Hell: A Survey of Science Fiction*, Penguin, London, 1960, p. 12.

18 Oleg Bestsennyy, Greg Gilbert, Alex Harris and Jennifer Rost, 'Telehealth: A Quarter-trillion-dollar Post-COVID-19 Reality?', *McKinsey & Company*, 9 July 2021, www.mckinsey.com/industries/healthcare-systems-and-services/our-insights/telehealth-a-quarter-trillion-dollar-post-covid-19-reality

19 See, for example, Thomas Hooven, 'Why Are We Dragging our Feet When More Innovation in Health Care Will Save Lives?' in *The Conversation*, 21 April 2021, https://theconversation.com/why-are-we-dragging-our-feet-when-more-automation-in-health-care-will-save-lives-75591

20 See, for example, Amy Jeter Hansen, 'Doctor's Reassurance Can Make Patients Feel Better, Study Finds', *Scope* (Stanford Medicine), 5 September 2018, https://scopeblog.stanford.edu/2018/09/05/doctors-reassurance-can-make-patients-feel-better-study-finds/; Barry D. Silverman, 'Physician Behavior and Bedside Manners: The Influence of William Osler and the Johns Hopkins School of Medicine', *Proceedings of the Baylor University Medical Center*, vol. 25, no. 1, 2012, pp. 58–61. https://doi.org/10.1080/08998280.2012.11928784

21 Anuja Vaidya, '51% of Clinicians Worry That Telehealth Hinders Ability to Show Empathy', *mHealth Intelligence*, 15 March 2022, https://mhealthintelligence.com/news/51-of-clinicians-worry-that-telehealth-hinders-ability-to-show-empathy

22 Paul Gardner, 'The Imminent Crisis of Mind', *Arena Online*, 27 January 2022, https://arena.org.au/the-imminent-crisis-of-mind/

23 Robert DiNapoli, 'Who Are You?', *Arena Online*, December 2021, https://arena.org.au/who-are-you/

24 Rosalind Picard, *Affective Computing*, MIT Press, Cambridge, Massachusetts, 1997, pp. 93–94.

25 Delphine Caruelle et al., 'Affective Computing in Marketing: Practical Implications and Research Opportunities Afforded by Emotionally Intelligent Machines', *Marketing Letters*, vol. 33, no. 1, 2022, pp. 163–69, https://doi.org/10.1007/s11002-021-09609-0

26 Adnan Khashman, 'Emotional System for Military Target Identification', Conference Paper, 2009, file:///C:/Users/Admin/Downloads/MP-IST-087-18-1.pdf

27 Jeanette Winterson, *12 Bytes: How We Got Here; Where We Might Go Next*, Vintage Books, New York, 2021, p. 155.

28 Sherry Turkle, op. cit., p. 55.

29 Jeanette Winterson, op. cit., p. 161.

30 Toby Walsh, 'Autonomous Weapons Open Letter: AI & Robotics Researchers', *Future of Life Institute*, 9 February 2016, https://futureoflife.org/2016/02/09/open-letter-autonomous-weapons-ai-robotics/

31 Toby Walsh, '*Eye in the Sky* Movie Gives a Real Insight into the Future of Warfare', *The Conversation*, 25 March 2016, https://theconversation.com/eye-in-the-sky-movie-gives-a-real-insight-into-the-future-of-warfare-56684

32 'General, your tank is a powerful vehicle' in Bertolt Brecht, *Poems 1913–1956*, Routledge, London, 1998, p. 289.

33 See, for example, Wayne Chappelle et al., 'An Analysis of Post-traumatic Stress Symptoms in United States Air Force Drone Operators', *Journal of Anxiety Disorders*, vol. 28, no. 5, June 2014, pp. 480–87, https://doi.org/10.1016/j.janxdis.2014.05.003

34 Martin Amis, *Visiting Mrs Nabokov and Other Excursions*, Penguin, London, 1993, p. 13.

35 Dan Sabbagh, 'Killer Drones: How Many Are There and Who Do They Target?', *The Guardian*, 18 November 2019, www.theguardian.com/news/2019/nov/18/killer-drones-how-many-uav-predator-reaper

36 Quoted in Tom Huhn (ed.), *The Cambridge Companion to Adorno*, Cambridge University Press, Cambridge, 2004, p. 101.

Chapter 4: Hacking Humanity

1   Terry Eagleton, 'Pretty Much Like Ourselves', *London Review of Books*, 4 September 1997, www.lrb.co.uk/the-paper/v19/n17/terry-eagleton/pretty-much-like-ourselves

2   John Carey (ed.), *The Faber Book of Utopias*, Faber and Faber, London, 1999, pp. 47–49.

3   Ray Kurzweil, 'Superintelligence and Singularity', in *Science Fiction and Philosophy*, John Wiley & Sons, Inc., New York, 2016, pp. 146–70.

4   Bob Doede, 'Transhumanism, Technology, and the Future: Posthumanity Emerging or Sub-humanity Descending?', *Appraisal*, vol. 7, no. 3, 2009, p. 39.

5   Mary Shelley, *Frankenstein; or, the Modern Prometheus*, Vintage Books, London, 2007 (first published 1818).

6   Max Weber, *The Sociology of Religion*, Beacon Press, Boston, 1993 (first published 1920).

7   Mark Blitz, 'Understanding Heidegger on Technology', *New Atlantis*, Winter 2014, www.thenewatlantis.com/publications/understanding-heidegger-on-technology

8   Paul Thagard, 'Cognitive Science', *The Stanford Encyclopedia of Philosophy (Fall 2008 Edition)*, https://plato.stanford.edu/archives/fall2008/entries/cognitive-science/

9   Ibid.

10 Bob Doede, 'Human Nature, Technology and Mind-Uploading', *Patheos*, 21 September 2018, www.patheos.com/blogs/humanflourishing/2018/09/2018-8-8-robert-doede-human-nature-technology-and-mind-uploading/

11 Chistof Koch, 'Will Machines Ever Become Conscious?', *Scientific American*, 1 December 2019, www.scientificamerican.com/article/will-machines-ever-become-conscious/

12 Raymond Tallis, 'About Aboutness', *Philosophy Now*, June/July 2019, https://philosophynow.org/issues/132/About_Aboutness

13 See, for example, Yuval Noah Harari, *Homo Deus: A Brief History of Tomorrow*, Vintage Books, New York, 2017, and David Christian, *Origin Story: A Big History of Everything*, Penguin, London, 2018.

14 Geoff Sharp, 'From Here to Eternity? Part 2', *Arena Magazine*, no. 89, 2007, p. 39.

15 Mary Midgley, 'Biotechnology and Monstrosity: Why We Should Pay Attention to the "Yuk Factor"', *The Hastings Center Report*, vol. 30, no. 5, 2000, pp. 7–15.

16 James Bridle, *New Dark Age: Technology and the End of the Future*, Verso Books, New York, 2019, pp. 75–77.

17 Sheila Jasanoff, op. cit., p. 3.

18 Ibid., pp. 4–6.

19 Harry Collins, *Artifictional Intelligence: Against Humanity's Surrender to Computers*, Polity Press, Cambridge, 2018.

20 Raymond Tallis, 'Aping Mankind? Neuromania, Darwinitis and the Misrepresentation of Humanity' (lecture), 6 December 2012, www.youtube.com/watch?v=U5baL9oh430

21 For an excellent account of this phenomenon, sees Adam Curtis's series of documentary films, *The Trap*, 2007, available on YouTube.

22   Zac Rogers, 'The Coronavirus Pandemic is Boosting the Big Tech Transformation to Warp Speed', *The Conversation*, 29 May 2020, https://theconversation.com/the-corona virus-pandemic-is-boosting-the-big-tech-transformation-to-warp-speed-138537

23   Philip Mirowski and Edward Nik-Khah, *The Knowledge We Have Lost in Information: The History of Information in Modern Economics*, Oxford University Press, Oxford, 2017.

24   Bob Doede, *Transhumanism*, op. cit.

Chapter 5: Off-Target Effects

1   Aaron Bastani, *Fully Automated Luxury Communism: A Manifesto*, Verso Books, New York, 2020, p. 152.

2   Jon Cohen, 'China's CRISPR Push in Animals Promises Better Meat, Novel Therapies, and Pig Organs for People', *Science.org*, 31 July 2019, www.science. org/content/article/china-s-crispr-push-animals-promises-better-meat-novel-therapies-and-pig-organs-people

3   Erik Parens and Josephine Johnston, *Human Flourishing in an Age of Gene Editing*, Oxford University Press, Oxford, 2019, p. 2.

4   Stephen Asma, *On Monsters: An Unnatural History of Our Worst Fears*, Oxford University Press, Oxford, 2009, p. 269.

5   Steven Rose, 'The Limits to Science', *Jacobin*, 5 February 2018, www.jacobinmag. com/2018/05/science-ideology-ethics-inequality-genetics

6   Rachel Swaby, 'Scientists Create First Self-replicating Synthetic Life', *Wired*, 20 May 2010, www.wired.com/2010/05/scientists-create-first-self-replicating-synthetic-life-2/

7   Alex Valentine et al., 'The "Atom-Splitting" Moment of Synthetic Biology: Nuclear Physics and Synthetic Biology Share Common Features', *EMBO Reports*, vol. 13, no. 8, 2012, pp. 677–79.

8   Kate Cregan, 'Cleaving at the Root', *Arena Magazine*, April–May 2002.

9   See, for example, Tim Maloney, 'Galileo Goes to Annapolis', *The Washington Post*, 8 January 2006, www.washingtonpost.com/archive/2006/01/08/galileo-goes-to-annapolis/df750c18-1b19-4792-8867-fd3395b8285d/

10   Sheila Jasanoff, op. cit., p. 89.

11   Simon Cooper, 'The Small Matter of Our Humanity', *Arena Magazine*, June–July 2002, pp. 34–38.

12   Sheila Jasanoff, op. cit., p. 15.

13   Ibid., p. 92.

14   Ibid., p. 15; Simon Cooper, 'Techno-science and the Post-human Condition', *Arena Magazine*, April–May 2019, pp. 47–51.

15   Sheila Jasanoff, op. cit., p. 174.

16   Kate Cregan, 'On the Frontier of the Bio-tech Boom', *Arena Magazine*, August–September 2001, pp. 6–8.

17   Sheila Jasanoff, op. cit., p. 62.

18   Quoted in Jim Holt, 'Measure for Measure', *The New Yorker*, 24 January 2005, www.newyorker.com/magazine/2005/01/24/measure-for-measure-5

19   Quoted in Kate Manne, 'Reconsider the Lobster: Jordan Peterson's Failed Antidote for Toxic Masculine Despair', *Times Literary Supplement*, 25 May 2018.

20 Raymond Tallis, *Aping Mankind? Neuromania, Darwinitis and the Misrepresentation of Humanity*, Routledge, London, 2014.

21 Anthony Smith, 'The Human Genome Project', *Arena Magazine*, February–March 1993.

22 For a study of the 'red market', see Scott Carney, *The Red Market: On the Trail of the World's Organ Brokers, Bone Thieves, Blood Farmers, and Child Traffickers*, William Morrow, New York, 2011.

23 For an example of such a company, see The Fertility Institutes, www.fertility-docs.com/. See also Editorial, 'The Alarming Rise of Complex Genetic Testing in Human Embryo Selection', *Nature*, 21 March 2022.

24 Quoted by the Centre for Genetics and Society, *Biopolitical Times*, 22 October 2007: 'If you really are stupid, I would call that a disease ... The lower 10 percent who really have difficulty, even in elementary school, what's the cause of it? A lot of people would like to say, "Well, poverty, things like that." It probably isn't. So I'd like to get rid of that, to help the lower 10 percent.' (As for how 'soft' such eugenics would be, readers are invited to cross-reference the comments Watson has made about the intellectually ungifted with those he has made about black IQs and formulate their own conclusions.)

25 Kate Cregan, op. cit.

26 Peter Singer, *Practical Ethics*, Cambridge University Press, Cambridge, 1980, p. 151.

27 Mary Midgley, op. cit., p. 9.

28 Michael J. Sandel, 'The Case Against Perfection', *The Atlantic*, April 2004, www.theatlantic.com/magazine/archive/2004/04/the-case-against-perfection/302927/

29 Ibid.

30 Alison Caddick, 'Liberal Post-humanism? Peter Singer and the Genetic Manipulation of Intelligence', *Arena Magazine*, October–November 2000.

31 Mary Midgley, op. cit, p. 12.

Chapter 6: Project Cyborg

1 Karl Miller, *Doubles: Studies in Literary History*, Oxford University Press, Oxford, 1987, p. 135.

2 The California-based company Ambrosia – named for the food that was said to make the Greek gods immortal – was one of the first to offer that product. Its CEO, Jesse Karmazin, claimed that the treatment came 'pretty close' to immortality. See Peter Ward, 'The Price of Silicon Valley's Obsession with Immortality', *Big Think*, 21 April 2022, https://bigthink.com/health/immortality-race-to-live-forever/

3 Aubrey de Grey (in interview), 'How to End Ageing with Aubrey de Grey', *Heights*, 15 April 2021, www.yourheights.com/blog/longevity/how-to-end-ageing-with-aubrey-de-grey

4 See, for example, the profile of Serge Faguet in Stefanie Marsh, 'Extreme Biohacking: The Tech Guru Who Spent $250,000 Trying to Live Forever', *The Guardian*, 21 September 2018, www.theguardian.com/science/2018/sep/21/extreme-biohacking-tech-guru-who-spent-250000-trying-to-live-for-ever-serge-faguet

5   According to David Christian, these 'thresholds' are as follows: the creation of matter in the wake of the Big Bang; the formation of stars and galaxies; the emergence of chemical complexity; the formation of the Earth and solar system; the emergence of life on Earth; the emergence of *Homo sapiens*; the development of agriculture; and the dramatic and possibly catastrophic emergence of the modern world, or the Anthropocene.

6   For an excellent discussion of Harari's essentially biologistic outlook, see Nick Spencer, 'Sapiens, Maybe; Deus, No: The Problem with Yuval Noah Harari', *ABC Religion and Ethics*, 13 July 2020, www.abc.net.au/religion/the-problem-with-yuval-noah-harari/12451764

7   The Big History website is https://bhp-public.oerproject.com/

8   Manfred E. Clynes and Nathan S. Kline, 'Cyborgs and Space', Astronautics (reprinted in *The New York Times*), September 1960, https://archive.nytimes.com/www.nytimes.com/library/cyber/surf/022697surf-cyborg.html

9   The full text of the interview can be found at www.naturearteducation.org/R/Artikelen/Betrayal.htm

10   Joseph Lee, 'Cochlear Implantation, Enhancements, Transhumanism and Posthumanism: Some Human Questions', *Science and Engineering Ethics*, vol. 22, no. 1, 2016, pp. 67–92.

11   Stephen Asma, op. cit., p. 269.

12   Hazel Moir and Deborah Gleeson, 'Explainer: Evergreening and how Big Pharma Keeps Drug Prices High', *The Conversation*, 6 November 2014, https://theconversation.com/explainer-evergreening-and-how-big-pharma-keeps-drug-prices-high-33623

13   'There is no difference between saving lives and extending lives,' says Aubrey de Grey, 'because in both cases we are giving people the chance of more life.'

14   Margaret Osborne, 'Brain Implants Allow Paralyzed Man to Communicate Using His Thoughts', *Smithsonian Magazine*, 25 March 2022, www.smithsonianmag.com/smart-news/brain-implants-allow-paralyzed-man-to-communicate-180979817/explainer-evergreening-and-how-big-pharma-keeps-drug-prices-high-33623

15   Slavoj Žižek, 'Elon Musk's Desire to Control Our Minds Is Dehumanizing and Not What Is Needed in a Socially Distanced World', *RT.com*, 1 September 2020, www.rt.com/op-ed/499626-slavoj-zizek-elon-musk/

16   Quoted in Sarah Marsh, 'Neurotechnology, Elon Musk and the Goal of Human Enhancement', *The Guardian*, 1 January 2018, www.theguardian.com/technology/2018/jan/01/elon-musk-neurotechnology-human-enhancement-brain-computer-interfaces

17   Donna Haraway, *A Cyborg Manifesto: Science, Technology, and Socialist-Feminism in the Late Twentieth Century*, University of Minnesota Press, 2016 (first published 1984), p. 15.

18   Foucault, Michel, *The Order of Things: An Archaeology of the Human Sciences*, Routledge, London, 2002 (first published 1966), p. 422.

19   Judith Butler, *Gender Trouble: Feminism and the Subversion of Identity*, Routledge, London, 2006, p. 47.

Chapter 7: Some Sweet Oblivious Antidote

1   Aldous Huxley, *Brave New World*, Coradella Collegiate Bookshelf Editions, 2004 (first published 1932), p. 265.

2   Robert Bennett, 'Aldous Huxley Foresaw America's Pill-popping Addiction with Eerie Accuracy', *Lithub.com*, 21 March 2019, https://lithub.com/aldous-huxley-foresaw-americas-pill-popping-addiction-with-eerie-accuracy/

3   Doug Hendrie, 'An Antidepressant Is Now One of Australia's Most Commonly Prescribed Drugs – But Why?', *newsGP*, 1 December 2020, www1.racgp.org.au/newsgp/clinical/an-antidepressant-is-now-one-of-australia-s-most-c

4   Zalika Rizmal, 'Australians are Taking Antidepressants in Record Numbers and For Longer Than Ever, But Coming Off Them Can Be Frightening', *ABC Online*, 22 February 2022, www.abc.net.au/news/2022-02-22/australians-coming-off-antidepressants/100847462

5   Juliette Jowit, 'What Is Depression and Why Is It Rising?', *The Guardian*, 4 June 2018, www.theguardian.com/news/2018/jun/04/what-is-depression-and-why-is-it-rising; World Health Organization, 'COVID-19 Pandemic Triggers 25% Increase in Prevalence of Anxiety and Depression Worldwide', WHO, 2022, www.who.int/news/item/02-03-2022-covid-19-pandemic-triggers-25-increase-in-prevalence-of-anxiety-and-depression-worldwide

6   Institute for Health Metrics and Evaluation, 'New Global Burden of Disease Analyses Show Depression and Anxiety among the Top Causes of Health Loss Worldwide, and a Significant Increase due to the COVID-19 Pandemic', IHME, 8 October 2021, www.healthdata.org/acting-data/new-ihme-analyses-show-depression-and-anxiety-among-top-causes-health-burden-worldwide

7   Bandon H. Hidaka, 'Depression as a Disease of Modernity: Explanations for Increasing Prevalence', *Journal of Affective Disorders*, vol. 140, no. 3, 2011, pp. 205–14.

8   Karl Polanyi, *The Great Transformation: The Political and Economic Origins of Our Time*, Beacon Press, 2001 (first published 1944), Chapter 4.

9   David Harvey, *The Anti-Capitalist Chronicles*, Pluto Press, London, 2020, p. 19.

10   Christopher Lasch, *The Culture of Narcissism: American Life in An Age of Diminishing Expectations*, W. W. Norton, New York, 1991, p. 72.

11   For a general discussion of concept creep, see Gary Greenberg, *The Book of Woe: The DSM and the Unmaking of Psychiatry*, Scribe Publications, Melbourne, 2013.

12   Luigi Esposito and Fernando Perez, 'Neoliberalism and the Commodification of Mental Health', *Humanity & Society*, vol. 38, no. 4, 2014, pp. 414–42.

13   Mark Furlong, 'Loving Machines: Mental Health by Algorithm is Reshaping Care and Sociality', *Arena Online*, March 2022, https://arena.org.au/loving-machines-mental-health-by-algorithm-is-reshaping-care-and-sociality/

14   Quoted in Rina Raphael, 'This Therapy Robot that Lives in Facebook Messenger Wants to Treat Your Anxiety', *Fast Company*, 31 July 2017, www.fastcompany.com/40442761/can-this-therapy-robot-help-fix-the-depression-and-anxiety-epidemic-with-uplifting-gifs

15   Mark Furlong, op. cit.

16   Ibid.

17   Parker Crutchfield, '"Morality Pills" May Be the US's Best Shot at Ending the
      Coronavirus Pandemic, According to One Ethicist', *The Conversation*, 10 August
      2020, https://theconversation.com/morality-pills-may-be-the-uss-best-shot-at-
      ending-the-coronavirus-pandemic-according-to-one-ethicist-142601

18   Thomas Douglas, 'Moral Enhancement', *Journal of Applied Philosophy*, vol. 25,
      no. 3, 2008, pp. 228–45.

19   See, in particular, Ingmar Persson and Julian Savulescu, 'Moral Transhumanism',
      The Journal of Medicine and Philosophy, vol. 35, no. 6, 2010, pp. 656–69.

20   Molly Crockett et al., 'Serotonin Selectively Influences Moral Judgment and
      Behavior through Effects on Harm Aversion', *Proceedings of the National Academy
      of Sciences (PNAS)*, vol. 107, no. 40, 2010, pp. 17433–38.

21   Molly Crocket, 'Morality Pills: Reality or Science Fiction?', *The Guardian*,
      3 June 2013, www.theguardian.com/science/head-quarters/2014/jun/03/
      morality-pills-reality-or-science-fiction

Chapter 8: The Black Box Society

1   Carl Benedikt Frey and Michael A. Osborne, *The Future of Employment: How
      Susceptible Are Jobs to Computerisation?*, Oxford Martin School, 17 September 2013,
      www.oxfordmartin.ox.ac.uk/downloads/academic/The_Future_of_Employment.pdf

2   John Maynard Keynes, 'Economic Possibilities for Our Grandchildren', *Essays in
      Persuasion*, Macmillan, 1933, pp. 358–74.

3   Robert Nozick, *Anarchy, State and Utopia*, Blackwell, New Jersey, 1974, p. 42,
      https://archive.org/details/anarchystateutop00nozi/page/42/mode/2up

4   Felipe De Brigard, 'If You Like It, Does It Matter if It's Real?', *Philosophical
      Psychology*, vol. 23, no. 1, February 2010, pp. 43–57.

5   William Davies, *The Happiness Industry*, op. cit., loc. 353.

6   John Danaher, 'Will Life Be Worth Living in a World Without Work?
      Technological Unemployment and the Meaning of Life', *Science and Engineering
      Ethics*, vol. 23, no. 1, 2017, pp. 41–64.

7   Matthew B. Crawford, 'Shop Class as Soul Craft: The Case for the Manual
      Trades', *The New Atlantis*, Summer 2006, www.thenewatlantis.com/publications/
      shop-class-as-soulcraft

8   See 'Hazards of Prophecy: The Failures of Imagination' in Arthur C. Clark,
      Profiles of the Future: An Inquiry into the Limits of the Possible (revised edition),
      Harper & Row, New York, 1973, pp. 12–21.

9   Matthew B. Crawford, 'Virtual Reality as Moral Ideal', *The New Atlantis*, Winter
      2015, www.thenewatlantis.com/publications/virtual-reality-as-moral-ideal

10   Gavin Mueller, *Breaking Things at Work: The Luddites Were Right about Why You
      Hate Your Job*, Verso Books, New York, p. 25.

11   Ibid., p. 34.

12   Shannon Liao, 'Amazon Warehouse Workers Skip Bathroom Breaks to Keep Their
      Jobs, Says Report', *The Verge*, 17 April 2018, www.theverge.com/2018/4/16/17243
      026/amazon-warehouse-jobs-worker-conditions-bathroom-breaks

13   Michael Sainato, 'Amazon Could Run Out of Workers in US in Two Years,
      Internal Memo Suggests', *The Guardian*, 22 June 2022, www.theguardian.com/
      technology/2022/jun/22/amazon-workers-shortage-leaked-memo-warehouse

14   Will Davies, 'The Holiday of Exchange Value', *Political Economy Research Centre Blog*, 7 April 2020, www.perc.org.uk/project_posts/the-holiday-of-exchange-value/

15   Jonathan Kingsley et al., 'Experiences of Gardening During the Early Stages of the COVID-19 Pandemic', *Health & Place*, vol. 176, July 2022, p. 102854.

16   David Graeber, *Bullshit Jobs: A Theory*, Penguin, London, 2019.

Chapter 9: Breaking the Frame

1   Bruce Pascoe, *Dark Emu: Black Seeds: Agriculture of Accident?*, Magabala Books, Broome, 2014.

2   Peter Sutton and Keryn Walshe, *Farmers or Hunter-Gatherers? The Dark Emu Debate*, Melbourne University Press, Melbourne, 2021.

3   See, for example, Tony Thomas, 'Bruce Pascoe, Melbourne University's Former Aborigine', *Quadrant Online*, 27 October 2021, https://quadrant.org.au/?s=pascoe

4   The phrase 'underlying supremacism' was used by the Australian Senator and minister Penny Wong and is quoted in Stuart Rintoul, 'Debunking *Dark Emu*: Did the Publishing Phenomenon Get It Wrong?', *The Age* (*Good Weekend* supplement), 12 June 2021, www.theage.com.au/national/debunking-dark-emu-did-the-publishing-phenomenon-get-it-wrong-20210507-p57pyl.html

5   Clifford Geertz, 'The Transition to Humanity', *Horizons of Anthropology*, Aldine Pub. Co., London, 1977, p. 44.

6   David F. Channell, *A History of Technoscience: Erasing the Boundaries between Science and Technology*, Routledge, London, 2017, p. 2.

7   'Capital is dead labour, which, *vampire*-like, lives only by sucking living labour, and lives the more, the more labour it sucks.'

8   Elise Bohan, 'On Becoming Posthuman', *Griffith Review: The New Disrupters*, 64, Text Publishing, 2019, www.griffithreview.com/articles/becoming-posthuman-big-history/

9   Norman Geras, *Marx and Human Nature: Refutation of a Legend* (Kindle edition), Verso Books, New York, 2016 (first published 1983), loc. 1256.

10   Terry Eagleton, *Why Marx Was Right*, Yale University Press, New Haven, 2011, p. 81.

11   See, for example, Ann Pettifor, *The Case for a Green New Deal*, Verso Books, New York, 2019.

12   Damian Carrington, 'World Close to "Irreversible" Climate Breakdown, Warn Major Studies', *The Guardian*, 28 October 2022, www.theguardian.com/environment/2022/oct/27/world-close-to-irreversible-climate-breakdown-warn-major-studies

13   Mark Harris, 'Silicon Valley Billionaires Want to Geoengineer the World's Oceans', *New Scientist*, 1 September 2020, www.newscientist.com/article/mg24732980-500-silicon-valley-billionaires-want-to-geoengineer-the-worlds-oceans; Catherine Clifford, 'White House Is Pushing Ahead Research to Cool Earth by Reflecting Back Sunlight', *CNBC*, 13 October 2022, www.cnbc.com/2022/10/13/what-is-solar-geoengineering-sunlight-reflection-risks-and-benefits.html

14   Geoff Mann and Joel Wainwright, *Climate Leviathan: A Political Theory of Our Planetary Future*, Verso Books, New York, 2018.

15   Langdon Winner, op. cit., p. 34. Of the bomb, Winner writes: 'As long as it exists at all, its lethal properties demand that it be controlled by a centralised, rigidly hierarchical, chain of command closed to all influences that might make its workings unpredictable. [This is] a matter of practical necessity independent of any larger political system in which the bomb is embedded.'

Coda: The Human Scale

1   In *Earthrise: How Man First Saw the Earth* (2008), for example, UK historian Robert Poole suggests that it marked 'the moment when the sense of the space age flipped from what it meant for space to what it means for Earth'. Quoted in John Noble Wilford, 'On Hand for Space History, as Superpowers Spar', *The New York Times*, 13 July 2009, www.nytimes.com/2009/07/14/science/space/14mission.html

2   Carole Cadwalladr, 'Stewart Brand's *Whole Earth Catalog*, The Book that Changed the World', *The Guardian*, 5 May 2013, www.theguardian.com/books/2013/may/05/stewart-brand-whole-earth-catalog

3   Frédéric Neyrat, *The Unconstructable Earth: An Ecology of Separation*, Fordham University Press, New York, 2018, p. 35.

4   Primo Levi, *A Tranquil Star: Unpublished Stories*, Penguin, London, 2007, p. 156.

5   Hannah Arendt, *The Human Condition*, Doubleday Anchor Books, New York, 1959, p. 4.

6   Ibid., p. 157.

7   Hannah Arendt, *The Human Condition*, Doubleday Anchor Books, New York, 1959, p. 4.

Afterword

1   Katie Wickens, 'AI Image Generator Births the Horrific "First Cryptid of the Latent Space"', *PC Gamer*, 8 September 2022, www.pcgamer.com/ai-image-generaotr-loab-cryptid-supercomposite/

2   Devin Coldewey, 'A Terrifying AI-generated Woman is Lurking in the Abyss of Latent Space', *TechCrunch*, 14 September 2022, https://techcrunch.com/2022/09/13/loab-ai-generated-horror/